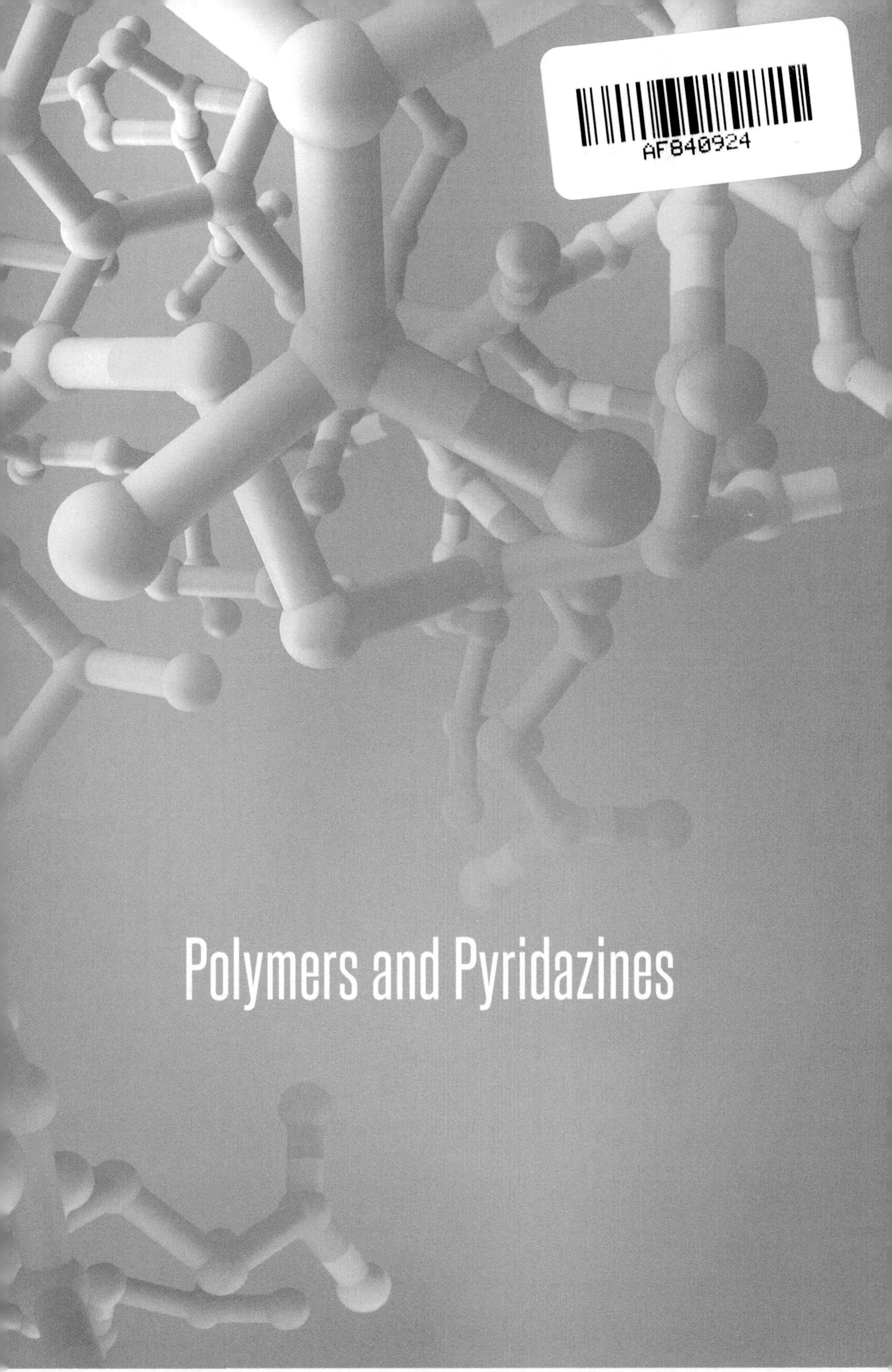

Polymers and Pyridazines

Polymers and Pyridazines

Péter Tétényi

JENNY STANFORD
PUBLISHING

Published by

Jenny Stanford Publishing Pte. Ltd.
Level 34, Centennial Tower
3 Temasek Avenue
Singapore 039190

Email: editorial@jennystanford.com
Web: www.jennystanford.com

British Library Cataloguing-in-Publication Data
A catalogue record for this book is available from the British Library.

Polymers and Pyridazines

ISBN 978-981-4800-47-1 (Paperback)
ISBN 978-0-429-29646-8 (eBook)

Contents

Preface

In the past couple of years, a few articles have been published on the solid-phase synthesis of pyridazine derivatives. These methods apply to intermediates weakly bound to polymers, as a result of which the ester bond is cleaved easily, either during the ring closure or right after it. We were going to develop a new strategy for polymer-supported synthesis of pyridazine derivatives, with much higher reaction rates, applying higher loading and much wider reaction conditions due to the more stable attachment. On the basis of our research, a fundamental breakthrough was achieved in solid-phase heterocyclic chemistry. Only monofunctional reagents were used for the attachment reaction to avoid the undesired double or polyreaction with the bifunctional hydrazine. The applied N^2-alkyl attachments proved to be stable during some nucleophilic substitution reactions of the attached substrate but could be easily cleaved by the end of the synthesis via attack of electrophiles (BI_3) or strong acids (TFA). Dynamic swelling measurements of the supported substrates simulated the behavior of the polymers in real reaction conditions. Optimization was done for each reaction step, and considerable acceleration of the initial reaction rates was usually reached under preparative conditions. Considerable widening of the applied reaction conditions was reached due to stability of the attachment. For example, it was stable within a pH range of 2–12 and up to a temperature of 170°C (approximate measured value). Our new linker with the phenol ether structural element in it was utilized as a multidetachable linker, and moreover, it was applied as a cosolvent in chemical reactions. Various comparisons of the areas of changing bands (FTIR) were carried out, with or without using control reactions, with or without applying electronic subtractions of the standardized spectra. The results were evaluated according to weight changes measured as well as according to elemental analyses.

Dr. Péter Tétényi

Department of Organic Chemistry, Pharmaceutical Faculty,
Semmelweis University, Hungary
2019

Acknowledgments

The author wants to thank Professor P. Mátyus for allowing heterocyclic chemical reactions on the polymer support. The author is also very thankful to his son, Péter, for his fast and precise help in formatting the figures.

No grants or funds were used for the experiments described in this book.

Chapter 1

Introduction

In the past couple of years, a few articles have been published on the solid-phase synthesis of pyridazine derivatives [1–9]. These methods apply to intermediates weakly bound to polymers, as a result of which the ester bond is cleaved easily, either during the ring closure or right after it. While these approaches have some advantages if someone wants to have quick success in a few steps while just only touching on solid-phase chemistry, these inhibit exploitation of the field (e.g., optimization of solid-phase synthetic steps is not possible because the product is immediately cleaved from the polymer). Moreover, many slow reactions are used in these methods [1–5, 8, 9] and it is disadvantageous to react or prepare sensitive substrates by reactions taking place over days. Since polymers are not dispersed generally in the whole volume of the reaction mixture, local concentrations of the polymer-supported substrates may be 5 to 20 times higher than that of the reagents (if an equimolar amount is used), depending on the swelling ability of the polymer in the solvent used. Low reaction rates are usually due to low concentrations of the reagents, because more vigorous conditions cannot be used due to the lability of the linker, and because physical properties (like swelling) of the polymers are considered neither by these methods nor by a related solid-phase strategy [10]. Supports of low loading (below 0.6 mmol/g) are used [1–10], but this is insufficient for preparative purposes. Multistage sequences of reaction steps with

Polymers and Pyridazines
Péter Tétényi

ISBN 978-981-4800-47-1 (Paperback), 978-0-429-29646-8 (eBook)
www.jennystanford.com

various nucleophiles present elegant approaches [3, 6, 8, 9], but only till the point where optimization is not needed. Otherwise, multistage sequences just cover the fact that simultaneous presence of multiple nucleophiles always results in a product mixture and that the optimum conditions have not been found yet.

In an article [9], a pyridazine ring attached to a Wang polymer is used, then a 6-aryl substituent is introduced by a Suzuki reaction, and finally the substrate is cleaved from the polymer by acids. Ring synthesis is not attempted [11], but still there are problems of regioselectivity during the Mitsunobu reaction (the Wang polymer reacts not only with the 3-OH but also with the 6-Cl of 6-chloropyridazin-3-ol). Nevertheless, this approach with the Wang polymer represents a real solid-phase synthesis with a similar pyridazine derivative we were going to apply.

The problem of analysis of the supported substrates is still not solved in these articles, since either infrared spectroscopy or weight increase/decrease during the reaction or various control reactions are used, but always only one of them for a given substrate. In the first article [3], serious misunderstanding of the infrared spectra is described, since the reported bands at 1382 and at 1139 cm^{-1} surely do not correspond to the presence of a polymer-supported pyridazine ring. Many authors dealing with polymer-supported synthesis do not recognize the importance of the presence of a C=O bond; without it, quantitative evaluation of the infrared changes is practically not possible. Elemental analyses of the supported substrates are seldom carried out, and these are not compared to other analytical results. Summarizing the previous literature data, it is concluded that no general method has been introduced for the solid-phase synthesis of pyridazine derivatives. Multiple checking methods have not been introduced, and further considerations must be made to overcome the danger of double/multiple reactions of the supported substrate. Our results and the new polymers and methods were based on our previous research published—among other papers—in three articles [12–14], but these results had to be developed and extended to a new field.

Chapter 2

Experiments

The reactions were carried out by magnetic stirring at about 300–800 rpm; the reaction mixtures were heated using a silicon oil bath electrically regulated within ±2°C. The infrared spectra were taken with a Perkin–Elmer 1600 Fourier-transform infrared spectroscopy (FTIR) spectrophotometer and were sent to a Perkin–Elmer data manager (PEDM) installed on an IBM-PC computer. These spectra were standardized using the blank, flat, smooth, and abex functions of the PEDM software and were evaluated according to the intensities of the changing peak, by the logarithm of the relative absorbances as well as by areas. The transmittance was set ranging from 100% to 1.5%, with a completely horizontal baseline for each spectrum. Areas of standardized spectra were only compared to each other, and other reference bands were needed only for the infrared difference spectra (IDS). The ^{1}H and ^{13}C nuclear magnetic resonance (NMR) spectra were recorded at a Bruker 200 MHz spectrometer and then on a Bruker 400 MHz spectrometer.

Swelling of the functionalized supports was measured in a pipette closed at the tip. The polymer was put with the solvent into the pipette immersed in an appropriate solvent at the proper temperature in a round-bottomed flask. Swelling results were evaluated after 5–10 minutes (dynamic swelling).

In this book, Ⓟ stands for 1% cross-linked styrene-divinylbenzene copolymer in a bead form.

Polymers and Pyridazines
Péter Tétényi

ISBN 978-981-4800-47-1 (Paperback), 978-0-429-29646-8 (eBook)
www.jennystanford.com

2.1 Preparation of 4,5-Dichloro-2-(polystyrylmethyl)-3(2*H*)-pyridazinone 5 (Attachment Reaction)

Equip a one-necked, round-bottomed flask of 500 cm^3 with a Dean–Stark trap with an inner volume of 10 cm^3 and with a Liebig condenser. Measure 10.32 g = 256 mmol sodium hydroxide into the flask and dissolve in 20 cm^3 of water. Add 105 cm^3 of butanol and 105 cm^3 of toluene while doing magnetic stirring and then add both 19.76 g = 120 mmol 4,5-dichloro-3(2*H*)-pyridazinone **3** and 6 g = 16.5 mmol hexadecyltrimethylammonium bromide (HTMAB) together. Stir the suspension for 10 minutes at 60°C. Add 20 g = 64 mmol chloromethyl polystyrene **4** and heat the reaction mixture at 130°C while stirring it. Add 30 cm^3 of anhydrous *N,N*-dimethylformamide (DMF) and continue stirring at 130°C–140°C by distilling out the aqueous solvent mixture from the reaction mixture. Continue distillation until you get a total of 23 cm^3 of aqueous phase (collected in a separate cylinder). Cool the reaction mixture to room temperature and filter it using a G3 glass suction filter. Wash the polymer with 30 cm^3 of water, 2 × 20 cm^3 of DMF, 20 cm^3 of 25% aqueous acetic acid, 10 cm^3 of methanol, 10 cm^3 of a methanol:toluene 1:1 mixture, 10 cm^3 of methanol, 10 cm^3 of a methanol:toluene 1:1 mixture, 2 × 10 cm^3 of methanol, and 2 × 5 cm^3 of diethyl ether. Dry the polymer in a drying apparatus at 100°C until you receive a constant mass.

Yield
30.2 g (100%) of a reddish-brown polymer with loading of 2.26 mmol/g

Infrared spectrum
νO–H, 3416 cm^{-1}; ν_{arom}C–H, 3020 cm^{-1}; ν_{aliph}C–H, 2922 and 2772 cm^{-1}; νC=O, 1600 cm^{-1}; ν_{arom}C=C, 1510 and 1450 cm^{-1}; γ_{arom}C–H, 858 and 812 cm^{-1}

IDS (product – starting material)
Appearing bands: νO–H, 3370 cm^{-1}; νC=O, 1662, 1630, and 1592 cm^{-1} Disappearing bands: βCH_2Cl, 1264 cm^{-1}; νC–Cl, 698 and 674 cm^{-1}

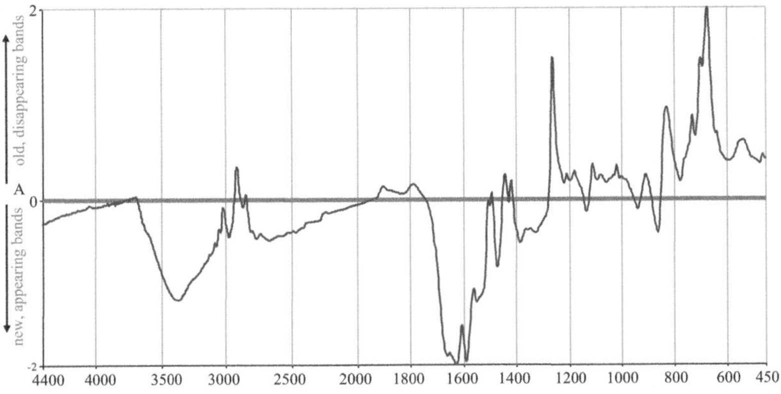

Time: 13:04:47 du. Date: 05.20. 2005 Prepared by Dr. Péter Tétényi

Organization: Department of Organic Chemistry Type of spectrum: standardized (Blank, Flat, Abex, Smooth) Resolution: 4 cm^{-1}

Instrument model: Perkin-Elmer FTIR-1600 Version ID: Report Builder, Rev. 1.10 TP 3355/A - TP 2753/A difference spectrum

2.2 Preparation of 4,5-Dichloro-2-(4-polystyryl(hydroxymethyl)phenyloxyethyl)-3(2*H*)-pyridazinone 7 (Attachment Reaction)

Equip a one-necked, round-bottomed flask of 100 cm^3 with a Dean–Stark trap with an inner volume of 10 cm^3 and a Liebig condenser. Measure 4.32 g = 0.11 mol sodium hydroxide into the flask and dissolve in 6 cm^3 of water. Add 40 cm^3 of butanol and 30 cm^3 of toluene while doing magnetic stirring and then add both 8.84 g = 44 mmol 4,5-dichloro-3(2*H*)-pyridazinone **3** and 1.27 g = 5 mmol HTMAB together. Stir the suspension for 30 minutes at 80°C. Add 7.25 g = 11.5 mmol 4-polystyryl(hydroxymethyl)phenyloxyethyl chloride **6** and add 60 cm^3 of DMF. Heat the reaction mixture to and keep it at 160°C. Distil out the aqueous solvent mixture from the reaction mixture. Continue distillation until you get a total of 8.1 cm^3 of aqueous phase (total distillation period 3.5 hours). Cool the reaction mixture to room temperature and filter using a G3 glass suction filter. Wash the polymer with 5 cm^3 of water, 2 × 5 cm^3 of DMF, 10 cm^3 of 25% aqueous hydrochloric acid, 5 cm^3 of methanol, 5 cm^3 of a methanol:toluene 1:1 mixture, 5 cm^3 of methanol, 5 cm^3 of a methanol:toluene 1:1 mixture, 2 × 5 cm^3 of methanol, and 2 × 5 cm^3

of diethyl ether. Dry the polymer in a drying apparatus at 50°C until you receive a constant mass.

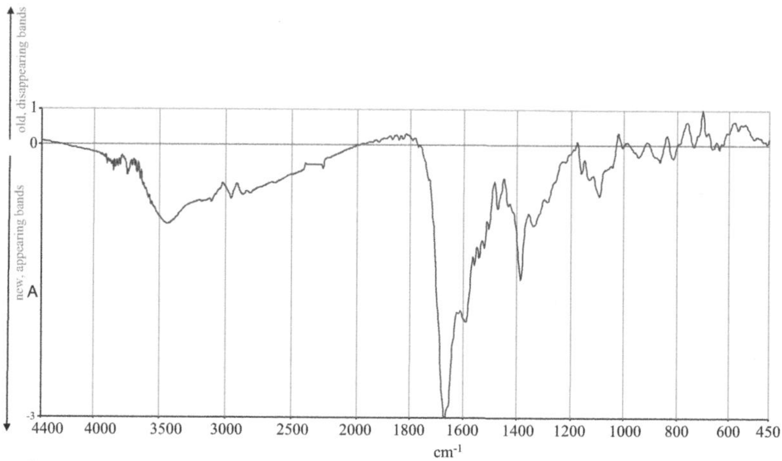

Time: 16:16:10. Date: 05.23.2006. TP 3421 - TP 3420 difference spectrum

Organization: Department of Organic Chemistry Type of spectrum: standardized (Blank, Flat, Abex, Smooth) Resolution: 4 cm^{-1}

Instrument model: Perkin-Elmer FTIR-1600 Version ID: Report Builder, Rev. 1.10 Prepared by Dr. Péter Tétényi

Yield

12.34 g (100%) of a brown polymer with loading of 1.20 mmol/g

Infrared spectrum

νO–H, 3420 cm^{-1}; ν_{arom}C–H, 3056 and 3022 cm^{-1}; ν_{aliph}C–H, 2920 and 2852 cm^{-1}; νC=O, 1668 and 1594 cm^{-1}; ν_{arom}C=C, 1492 and 1452 cm^{-1}; γ_{arom}C–H, 822, 758, and 700 cm^{-1}

IDS (product – starting material)

Appearing bands: νO–H, 3380 cm^{-1}; νC=O, 1668 and 1594 cm^{-1}

Disappearing bands: γ_{arom}C–C, 762 and 702 cm^{-1}

2.3 Preparation of 4-Chloro-5-iodo-2-(polystyrylmethyl)-3(2*H*)-pyridazinone 9 (Substitution Reaction)

Equip a one-necked, round-bottomed flask of 25 cm^3 with a Liebig condenser. Suspend 5 g = 7.05 mmol 4,5-dichloro-2-(polystyrylmethyl)-3(2*H*)-pyridazinone **5** in 30 cm^3 of anhydrous ethylene glycol (EG) while stirring. Add 23.8 g = 0.14 mol potassium iodide and 2.6 g = 7 mmol HTMAB to it. Stir the reaction mixture for 5 hours at 150°C. After the first hour, add 20 cm^3 of toluene to the reaction mixture and carry out water distillation using a Dean–Stark trap. Cool the reaction mixture to room temperature and filter using a G3 glass suction filter. Wash the polymer with 5 cm^3 of water, 2 × 5 cm^3 of DMF, 5 cm^3 of water, 5 cm^3 of methanol, 5 cm^3 of a methanol:toluene 1:1 mixture, 5 cm^3 of methanol, 5 cm^3 of a methanol:toluene 1:1 mixture, 2 × 5 cm^3 of methanol, and 2 × 5 cm^3 of diethyl ether. Dry the polymer in a drying apparatus at 50°C until you receive a constant mass.

Yield

5.33 g of a golden brown polymer (yield 98.4% according to IR analysis) with loading of 1.30 mmol/g

Infrared spectrum

νO–H, 3422 cm^{-1}; ν_{arom}C–H, 3020 cm^{-1}; ν_{aliph}C–H, 2920 and 2854 cm^{-1}; νC=O, 1600 cm^{-1}; ν_{arom}C=C, 1510 and 1452 cm^{-1}; γ_{arom}C–H, 812 cm^{-1}

IDS (product – starting material)

Appearing bands: νO–H, 3448 cm^{-1}; νC–C, 1084 cm^{-1}
Disappearing bands: νC=O, 1690 cm^{-1}; ν_{arom}C–C, 1582 cm^{-1}

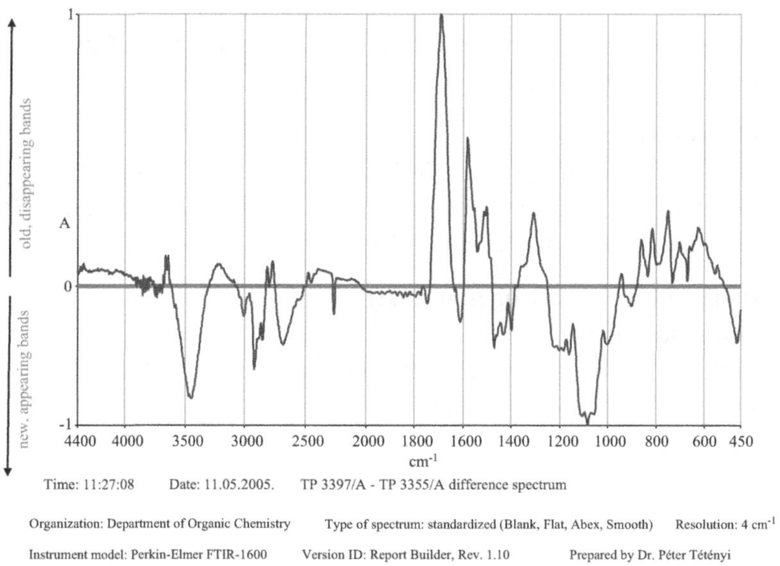

2.4 Preparation of 4-Chloro-5-iodo-2-(4-polystyryl(hydroxymethyl)phenyloxyethyl)-3(2*H*)-pyridazinone 10 (Substitution Reaction)

Equip a one-necked, round-bottomed flask of 50 cm^3 with a Liebig condenser and with a Dean–Stark trap. Suspend 2 g = 3.54 mmol 4,5-dichloro-2-(4-polystyryl(hydroxymethyl)phenyloxyethyl)-3(*2H*)-pyridazinone **7** in 12 cm^3 of anhydrous EG While stirring. Add 9.52 g = 57.4 mmol potassium iodide and 0.52 g = 1.4 mmol HTMAB. Add 30 cm^3 of anhydrous toluene to the reaction mixture and heat it to 150°C. Stir the reaction mixture at 150°C–160°C for 3 hours with water distillation; the volume of the total distilled aqueous phase should be 0.8 cm^3. Cool the reaction mixture to room temperature and filter using a G3 glass suction filter. Wash the polymer with 10 cm^3 of water, 2 × 5 cm^3 of DMF, 5 cm^3 of water, 5 cm^3 of methanol, 5 cm^3 of a methanol:toluene 1:1 mixture, 5 cm^3 of methanol, 5 cm^3 of a methanol:toluene 1:1 mixture, 2 × 5 cm^3 of methanol, and 2 × 5 cm^3 of diethyl ether. Dry the polymer in a drying apparatus at 50°C until you receive a constant mass.

Yield

1.45 g a golden brown polymer (yield 62.5% according to IR analysis) with loading of 1.10 mmol/g

Infrared spectrum

νO–H, 3420 cm^{-1}; ν_{arom}C–H, 3056 and 3022 cm^{-1}; ν_{aliph}C–H, 2920 and 2850 cm^{-1}; νC=O, 1652 cm^{-1}; ν_{arom}C=C, 1594, 1492, and 1450 cm^{-1}; γ_{arom}C–H, 820, 758, and 700 cm^{-1}

IDS (product – starting material)

Appearing bands: –

Disappearing bands: νO–H, 3394 cm^{-1}; ν_{aliph}C–H, 2920 cm^{-1}; νC=O, 1662 cm^{-1}; ν_{arom}C–C, 1596 and 1494 cm^{-1}

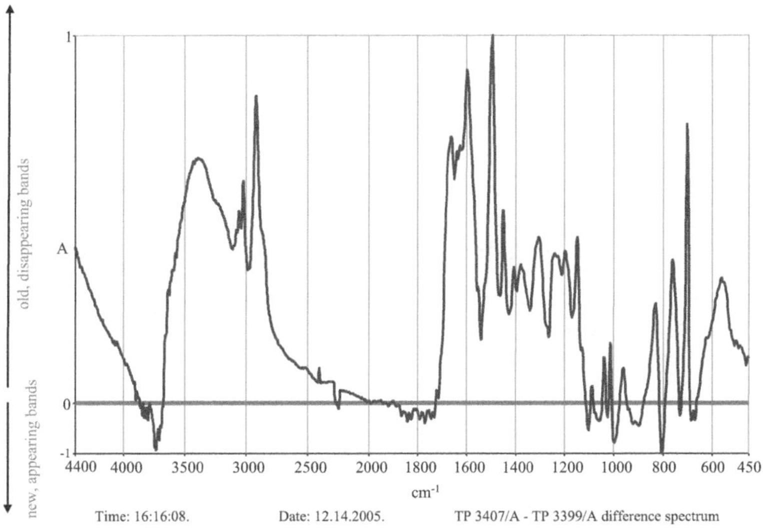

Time: 16:16:08. Date: 12.14.2005. TP 3407/A - TP 3399/A difference spectrum

Organization: Department of Organic Chemistry Type of spectrum: standardized (Blank, Flat, Abex, Smooth)

Instrument model: Perkin - Elmer FTIR - 1600 Version ID: Report Builder, Rev. 1.10 Prepared by Dr. Péter Tétényi

Resolution: 4 cm^{-1}

2.5 Preparation of 4-Chloro-5-pyrrolidino-2-(polystyrylmethyl)-3(2*H*)-pyridazinone 11 (Substitution Reaction)

Equip a one-necked, round-bottomed flask of 25 cm^3 with a Liebig condenser. Suspend 0.5 g = 0.60 mmol 4,5-dichloro-2-(polystyrylmethyl)-3(*2H*)-pyridazinone **5** in the mixture of 5 cm^3 of butanol and 5 cm^3 of toluene. Add 4.56 cm^3 = 54.7 mmol pyrrolidine and 0.22 g = 0.59 mmol HTMAB. Stir the reaction mixture for 3 hours at 85°C. Cool the reaction mixture to room temperature amd filter using a G3 glass suction filter. Wash the polymer with 10 cm^3 of water, 5 cm^3 of methanol, 5 cm^3 of a methanol:toluene 1:1 mixture, 5 cm^3 of methanol, 5 cm^3 of a methanol:toluene 1:1 mixture, 2 × 5 cm^3 of methanol, and 2 × 5 cm^3 of diethyl ether. Dry the polymer in a drying apparatus at 50°C until you receive a constant mass.

Yield

0.45 g of a yellow polymer (yield 96% according to IR analysis) with loading of 1.22 mmol/g

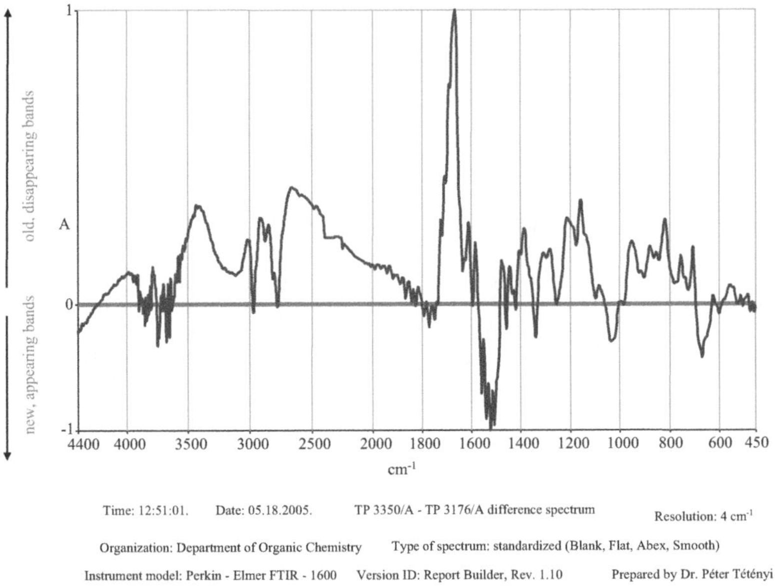

Infrared spectrum

νO–H, 3418 cm^{-1}; ν_{aliph}C–H, 2922, 2814, and 2776 cm^{-1}; νC=O, 1598 cm^{-1}; ν_{arom}C=C, 1508 and 1454 cm^{-1}; γ_{arom}C–H, 812 cm^{-1}

IDS (product – starting material)

Appearing bands: ν_{arom}C–N, 1540, 1522, and 1506 cm^{-1}
Disappearing bands: νO–H, 3430 cm^{-1}; νC=O, 1666 cm^{-1}; νC–C, 1156 cm^{-1}

2.6 Preparation of 4-Chloro-5-pyrrolidino-2-(4-polystyryl(hydroxymethyl)phenyloxyethyl)-3(2*H*)-pyridazinone 12 (Substitution Reaction)

Equip a one-necked, round-bottomed flask of 25 cm^3 with a Liebig condenser. Suspend 0.5 g = 0.65 mmol 4-chloro-5-iodo-2-(4-polystyryl(hydroxymethyl)phenyloxyethyl)-3(2*H*)-pyridazinone **10** in the mixture of 5 cm^3 of 1-butanol and 5 cm^3 of toluene. Add 3.0 cm^3 = 36 mmol pyrrolidine and 0.22 g = 0.59 mmol HTMAB. Stir the reaction mixture for 5 hours at 85°C. Cool the reaction mixture to room temperature and filter using a G3 glass suction filter. Wash the polymer with 10 cm^3 of water, 5 cm^3 of methanol, 5 cm^3 of a methanol:toluene 1:1 mixture, 5 cm^3 of methanol, 5 cm^3 of a methanol:toluene 1:1 mixture, 2 × 5 cm^3 of methanol, and 2 × 5 cm^3 of diethyl ether. Dry the polymer in a drying apparatus at 50°C until you receive a constant mass.

Yield

0.32 g of a brownish-red polymer (yield 95.3% according to infrared analysis) with loading of 1.16 mmol/g

Infrared spectrum

νO–H, 3384 cm^{-1}; ν_{aliph}C–H, 2920 cm^{-1}; νC=O, 1604 cm^{-1}; ν_{arom}C=C, 1508 and 1448 cm^{-1}; γ_{arom}C–H, 816 cm^{-1}

IDS (product – starting material)

Appearing bands: –

Disappearing bands: νO–H, 3446 cm^{-1}; ν_{aliph}C–H, 2918 cm^{-1}; νC=O, 1666 cm^{-1}; νC–O, 1388 cm^{-1}

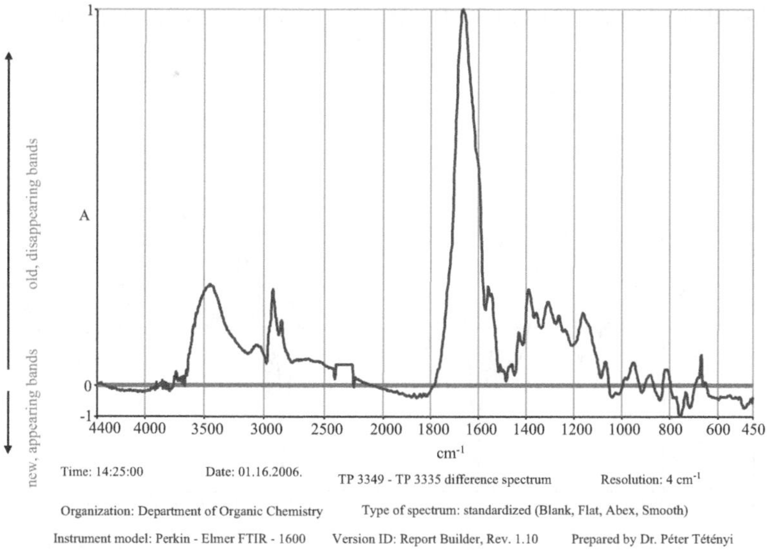

2.7 Preparation of 4-Chloro-5-[(2-sulfanylethyl)amino]-2-(polystyrylmethyl)-3(2*H*)-pyridazinone 13 (Substitution Reaction)

Equip a one-necked, round-bottomed flask of 25 cm^3 with a gas inlet adapter, with a Liebig condenser, and with a gas outlet adapter on top of the condenser. Evacuate the system and fill it with argon; repeat this process once. Suspend 0.5 g = 0.65 mmol 4-chloro-5-iodo-2-(polystyrylmethyl)-3(2*H*)-pyridazinone **9** in 6 cm^3 of anhydrous DMF. Add 0.63 g = 5.55 mmol 2-sulfanylethylammonium chloride and 0.92 g = 6.67 mmol anhydrous potassium carbonate and stir the reaction mixture for 3 hours at 90°C. Cool the reaction mixture to room temperature and add 5 cm^3 of water. Filter the reaction mixture through a G3 glass suction filter. Wash the polymer with 10 cm^3 of water, 10 cm^3 of 50% acetic acid (until you achieve a pH value of 4), 10 cm^3 of water, 5 cm^3 of methanol, 5 cm^3 of a methanol:toluene 1:1 mixture, 5 cm^3 of methanol, 5 cm^3 of a methanol:toluene 1:1 mixture, 2 × 5 cm^3 of methanol, and 2 × 5 cm^3 of diethyl ether. Dry the polymer in a drying apparatus at 50°C until you receive a constant mass.

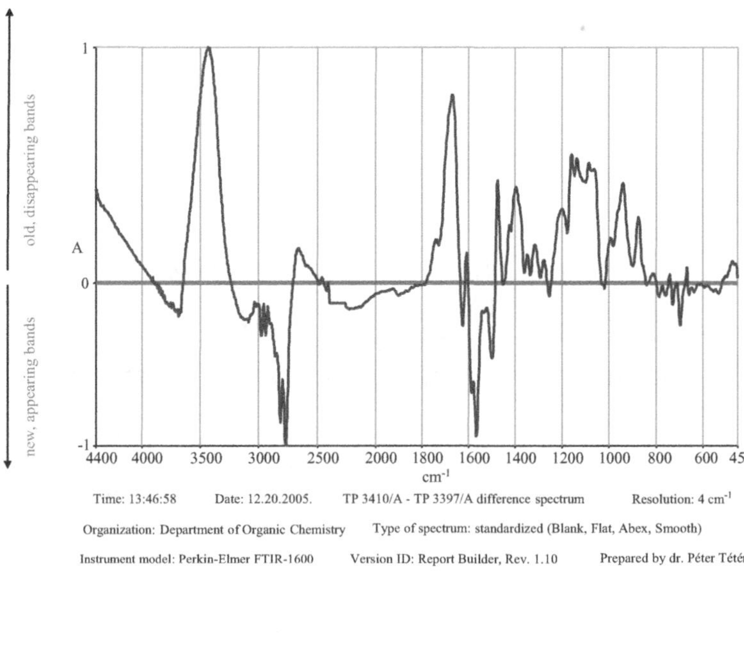

Yield
0.44 g of a yellowish-brown polymer (yield 94.5% according to infrared analysis) with loading of 1.29 mmol/g

Infrared spectrum
νO–H, 3406 cm^{-1}; ν_{arom}C–H, 3020 cm^{-1}; ν_{aliph}C–H, 2920, 2854, 2814, and 2770 cm^{-1}; νC=O, 1596 cm^{-1}; ν_{arom}C=C, 1510 and 1452 cm^{-1}; γ_{arom}C–H, 812 cm^{-1}

IDS (product – starting material)
Appearing bands: νC–H, 2814 and 2768 cm^{-1}; νC=O, 1586 cm^{-1}
Disappearing bands: νO–H, 3434 cm^{-1}; νC=O, 1670 cm^{-1}; δN–H, 1566 cm^{-1}; νC–C, 1158, 1136, and 1086 cm^{-1}

2.8 Preparation of 4-Chloro-5-[(2-sulfanylethyl)amino]-2-(4-polystyryl(hydroxymethyl)phenyloxyethyl)-3(2*H*)-pyridazinone 15 (Substitution Reaction)

Equip a one-necked, round-bottomed flask of 25 cm^3 with a gas inlet adapter, with a Liebig condenser, and with a gas outlet adapter on top of the condenser. Evacuate the system and fill it with argon; repeat this process once. Suspend 0.5 g = 0.23 mmol 4-chloro-5-iodo-2-(4-polystyryl(hydroxymethyl)phenyloxyethyl)-3(2*H*)-pyridazinone **10** in 5 cm^3 of anhydrous DMF. Add 0.13 g = 1.38 mmol 2-sulfanylethylammonium chloride and 0.19 g = 1.38 mmol anhydrous potassium carbonate and stir the reaction mixture for 3 hours at 90°C. Cool the reaction mixture to room temperature and add 10 cm^3 of water. Filter the reaction mixture through a G3 glass suction filter. Wash the polymer with 5 cm^3 of water, 5 cm^3 of 25% acetic acid (until you achieve a pH value of 4), 5 cm^3 of water, 5 cm^3 of methanol, 5 cm^3 of a methanol:toluene 1:1 mixture, 5 cm^3 of methanol, 5 cm^3 of a methanol:toluene 1:1 mixture, 2 × 5 cm^3 of methanol, and 2 × 5 cm^3 of diethyl ether. Dry the polymer in a drying apparatus at 50°C until you receive a constant mass.

Yield
0.46 g of a yellowish brown polymer (yield 99.5% according to infrared analysis) with loading of 0.57 mmol/g

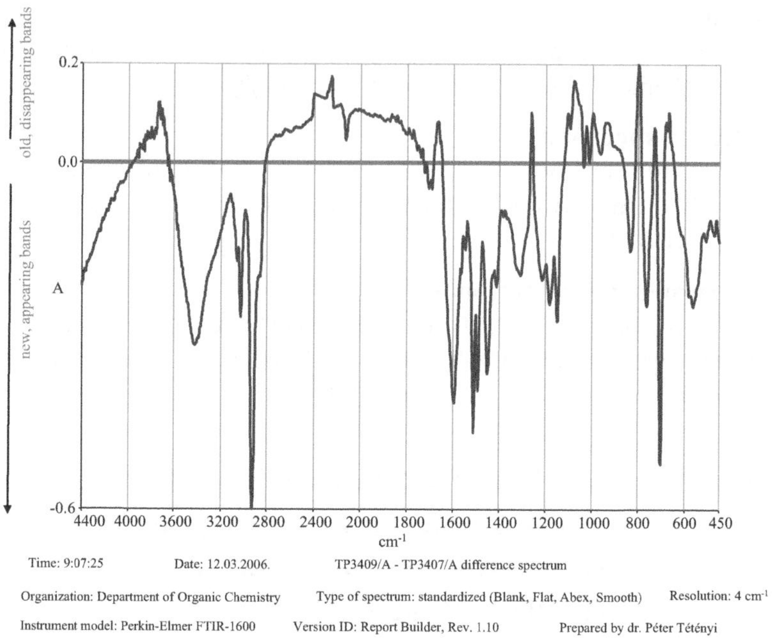

Infrared spectrum

νO–H and N–H, 3418 cm^{-1}; ν_{arom}C–H, 3056 and 3022 cm^{-1}; ν_{aliph}C–H, 2920 and 2852 cm^{-1}; νC=O, 1594 cm^{-1}; ν_{arom}C=C, 1508 and 1450 cm^{-1}; νC–O, 1220 cm^{-1}; γ_{arom}C–H, 822, 760, and 700 cm^{-1}

IDS (product – starting material)

Appearing bands: ν_{aliph}C–H, 2922 cm^{-1}; νC=O, 1596 cm^{-1}; νC–C, 1512 cm^{-1}; γ_{arom}C–H, 762 and 702 cm^{-1}

Disappearing bands: δN–H 1548 cm^{-1}; γ_{arom}C–H, 800 cm^{-1}

2.9 Preparation of 4-Chloro-5-hydrazino-2-(polystyrylmethyl)-3(2*H*)-pyridazinone 16 (Substitution Reaction)

Equip a one-necked, round-bottomed flask of 25 cm^3 with a Liebig condenser. Suspend 0.5 g = 0.65 mmol 4-chloro-5-iodo-2-(polystyrylmethyl)-3(2*H*)-pyridazinone **9** in 5 cm^3 of anhydrous acetonitrile (ACN) while stirring. Add 1.20 cm^3 of 24 mmol hydrazine hydrate to it and boil the reaction mixture for 3 hours at 90°C. Cool the reaction mixture to room temperature and filter using a G3 glass suction filter. Wash the polymer with 10 cm^3 of water, 5 cm^3 of methanol, 5 cm^3 of a methanol:toluene 1:1 mixture, 5 cm^3 of methanol, 5 cm^3 of a methanol:toluene 1:1 mixture, 2 × 5 cm^3 of methanol, and 2 × 5 cm^3 of diethyl ether. Dry the polymer in a drying apparatus at 50°C until you receive a constant mass.

Yield

0.44 g of an orange-colored polymer (yield 71.6% according to infrared analysis) with loading of 1.18 mmol/g

Infrared spectrum

νO–H, 3406 cm^{-1}; ν_{arom}C–H, 3018 cm^{-1}; ν_{aliph}C–H, 2922, 2854, 2814, and 2770 cm^{-1}; νC=O, 1600 cm^{-1}; ν_{arom}C=C, 1510 and 1450 cm^{-1}; γ_{arom}C–H, 854, 814, 760, and 702 cm^{-1}

IDS (product – starting material)

Appearing bands: –

Disappearing bands: νO–H, 3424 cm^{-1}; νC=O, 1668, 1626, and 1600 cm^{-1}

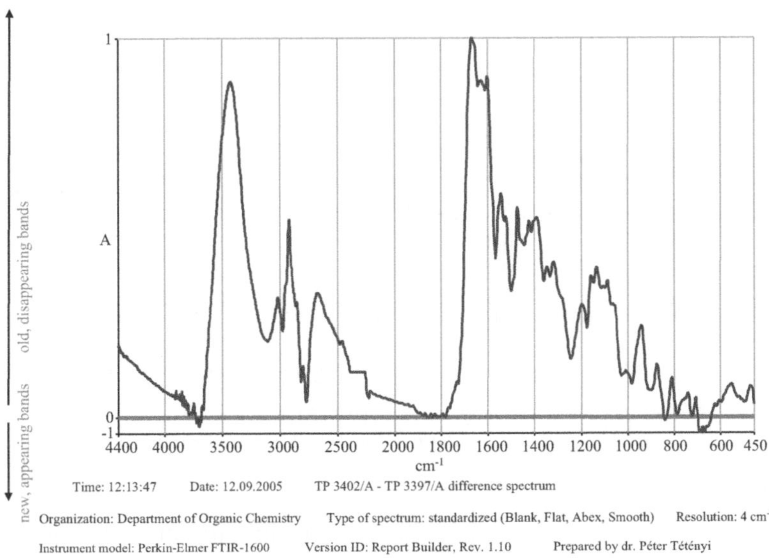

2.10 Preparation of 4-Chloro-5-hydrazino-2-(4-polystyryl(hydroxymethyl)phenyloxyethyl)-3(2*H*)-pyridazinone 17 (Substitution Reaction)

Equip a one-necked, round-bottomed flask of 25 cm^3 with a Liebig condenser. Suspend 8.85 g = 10.6 mmol 4,5-dichloro-2-(4-polystyryl(hydroxymethyl)phenyloxyethyl)-3(2*H*)-pyridazinone **10** in 50 cm^3 of anhydrous ACN while stirring. Add 13.30 cm^3 = 0.27 mol hydrazine hydrate to it and boil the reaction mixture for 4.5 hours at 90°C. Cool the reaction mixture to room temperature and filter using a G3 glass suction filter. Wash the polymer with 10 cm^3 of water, 5 cm^3 of methanol, 5 cm^3 of a methanol:toluene 1:1 mixture, 5 cm^3 of methanol, 5 cm^3 of a methanol:toluene 1:1 mixture, 2 × 5 cm^3 of methanol, and 2 × 5 cm^3 of diethyl ether. Dry the polymer in a drying apparatus at 50°C until you receive a constant mass.

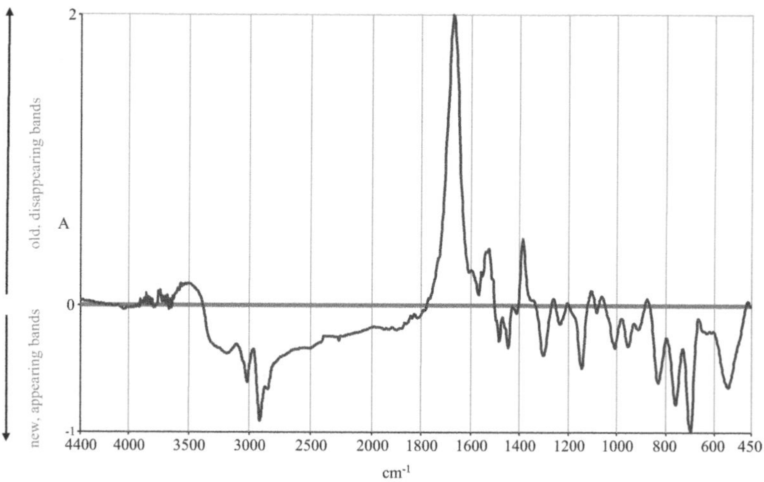

Time: 13:33:23 Date: 08.23.2006 TP 3446 - TP 3421difference spectrum Resolution: 4 cm^{-1}

Organization: Department of Organic Chemistry Type of spectrum: standardized (Blank, Flat, Abex, Smooth)

Instrument model: Perkin-Elmer FTIR-1600 Version ID: Report Builder, Rev. 1.10 Prepared by dr. Péter Tétényi

Yield

5.4 g of a red polymer (yield 88.2% according to infrared analysis) with loading of 1.08 mmol/g

Infrared spectrum
νO–H and N–H, 3422 cm^{-1}; ν_{arom}C–H, 3022 cm^{-1}; ν_{aliph}C–H, 2922 and 2852 cm^{-1}; νC=O, 1598 cm^{-1}; ν_{arom}C=C, 1492 and 1450 cm^{-1}; νC–O, 1222 cm^{-1}; γ_{arom}C–H, 822, 760, and 700 cm^{-1}

IDS (product – starting material)
Appearing bands: ν_{arom}C–H, 3016 cm^{-1}; ν_{aliph}C–H, 2912 cm^{-1}; δN–H, 1526 cm^{-1}; νC–N, 1302 cm^{-1}; γ_{arom}C–H, 830, 758, and 696 cm^{-1}
Disappearing bands: νC=O, 1672 cm^{-1}

2.11 Preparation of 4-Chloro-2-(polystyrylmethyl)-3(2*H*)-pyridazinone 18 (Elimination Reaction)

Equip a one-necked, round-bottomed flask of 50 cm^3 with a Dean-Stark trap and a Liebig condenser. Add 0.2 g = 0.55 mmol HTMAB and 1.7 g = 10 mmol copper(II) chloride·2H_2O and then add 10 cm^3 of anhydrous DMF and 20 cm^3 of anhydrous toluene. Add 0.5 g = 0.92 mmol 4-chloro-5-hydrazino-2-(polystyrylmethyl)-3(2*H*)-pyridazinone **16** to it and heat the reaction mixture to 160°C. Boil the reaction mixture for 1 hour at 160°C with distillation of the water. The volume of the distilled aqueous phase should be 1 cm^3. Cool the reaction mixture to room temperature and filter using a G3 glass suction filter. Wash the polymer with 10 cm^3 of water, 5 cm^3 of methanol, 5 cm^3 of a methanol:toluene 1:1 mixture, 5 cm^3 of methanol, 5 cm^3 of a methanol:toluene 1:1 mixture, 2 × 5 cm^3 of methanol, and 2 × 5 cm^3 of diethyl ether. Dry the polymer in a drying apparatus at 50°C until you receive a constant mass.

Yield
0.56 g of a brown polymer (yield 62.5% according to infrared analysis) with loading of 1.22 mmol/g

Infrared spectrum
νO–H, 3424 cm^{-1}; ν_{arom}C–H, 3020 cm^{-1}; ν_{aliph}C–H, 2922 cm^{-1}; $\nu_{carboxylate}$C=O, 1606 cm^{-1}; ν_{arom}C=C, 1510 and 1454 cm^{-1}; γ_{arom}C–H, 816 and 702 cm^{-1}

IDS (product – starting material)
Appearing bands: νO–H, 3478 cm^{-1}; $\nu_{saturated\ lactame}$C=O, 1692 cm^{-1}

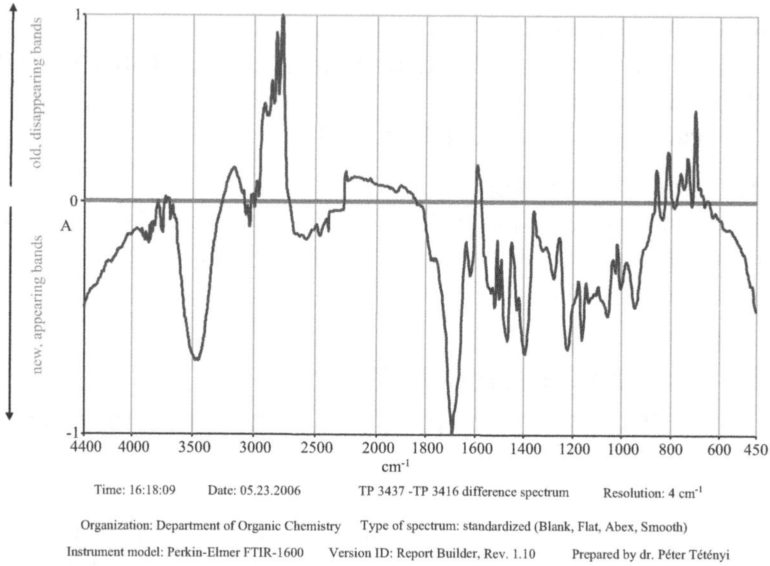

Disappearing bands: ν_{aliph}C–H, 2912, 2852, and 2814 cm^{-1}; $\nu_{saturated\ lactame}$C=O, 1692 cm^{-1}

2.12 Preparation of 4-Chloro-2-(4-polystyryl(hydroxymethyl)phenyloxyethyl)-3(2*H*)-pyridazinone 19 (Elimination Reaction)

Equip a one-necked, round-bottomed flask of 100 cm^3 with a Dean–Stark trap and a Liebig condenser. Add 2 g = 5.5 mmol HTMAB and 34 g = 0.2 mol copper(II) chloride·2H_2O and then add 30 cm^3 of anhydrous DMF and 40 cm^3 of anhydrous toluene. Add 5.4 g = 5.7 mmol 4-chloro-5-hydrazino-2-(4-polystyryl(hydroxymethyl)phenyloxyethyl)-3(2*H*)-pyridazinone **17** while stirring strongly. Heat the reaction mixture to and keep it at 160°C for 2 hours. During that time, carry out distillation of the water. The volume of the distilled aqueous phase should be 7.3 cm^3 (98.9% of the theoretical amount). Cool the reaction mixture to room temperature and filter using a G3 glass suction filter. Wash the polymer with 10 cm^3 of water, 5 cm^3 of methanol, 5 cm^3 of a methanol:toluene 1:1 mixture, 5 cm^3 of methanol, 5 cm^3 of a methanol:toluene 1:1 mixture, 2 × 5 cm^3

of methanol, and 2 × 5 cm^3 of diethyl ether. Dry the polymer in a drying apparatus at 50°C until you receive a constant mass.

Yield

5.38 g of a brown polymer (yield 61.5% according to infrared analysis) with loading of 0.68 mmol/g

Infrared spectrum

νO–H, 3424 cm^{-1}; ν_{arom}C–H, 3054 and 3022 cm^{-1}; ν_{aliph}C–H, 2920 and 2854 cm^{-1}; νC=O, 1656 and 1598 cm^{-1}; ν_{arom}C=C, 1490 and 1450 cm^{-1}; γ_{arom}C–H, 824, 760, and 700 cm^{-1}

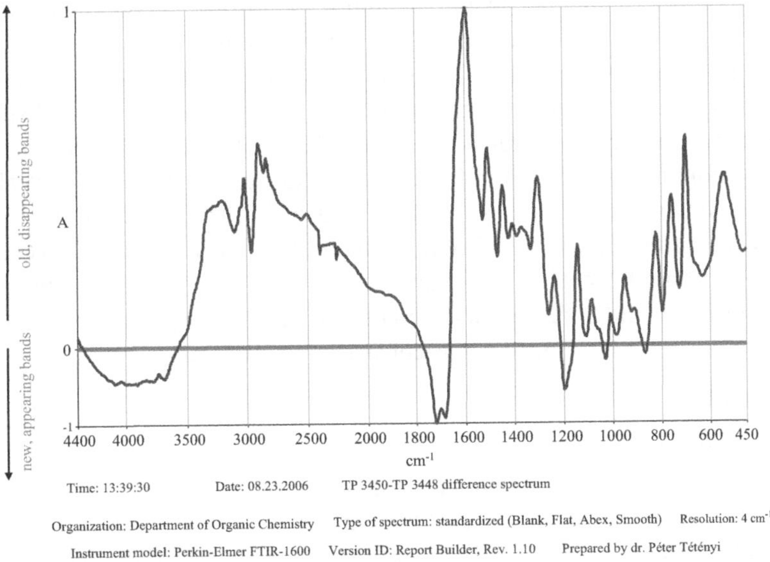

IDS (product – starting material)

Appearing bands: νC=O, 1678 cm^{-1}; νC–O, 1260 cm^{-1}; νC–C, 1196 cm^{-1}

Disappearing bands: νN–H, 3200 cm^{-1}; ν_{arom}C–H, 3018 cm^{-1}; ν_{aliph}C–H, 2902 and 2838 cm^{-1}; νC=O, 1594 cm^{-1}; γ_{arom}C–H, 756 and 698 cm^{-1}

The mother liquor after filtration was evaporated by reduced pressure, resulting in 2.18 g brownish-red raw crystals. This raw product was column-chromatographed twice by dichloromethane:butanol eluents in changing ratios, resulting in 0.34 g (19.2%) of cleaved hydroxy unsaturated lactame.

Infrared spectrum of the cleaved product
νO–H, 3422 cm^{-1}; ν_{arom}C–H, 3020 cm^{-1}; ν_{aliph}C–H, 2920 and 2852 cm^{-1}; νC=O, 1634 cm^{-1}; ν_{arom}C=C, 1470 cm^{-1}; γ_{arom}C–H, 722 cm^{-1}

2.13 Preparation of 4-Chloro-5-[(2-sulfanylethyl)amino]-3(2*H*)-pyridazinone 23 (Cleavage Reaction)

Equip a one-necked, round-bottomed flask of 25 cm^3 with a Liebig condenser. Suspend 0.44 g = 0.57 mmol 4-chloro-5-[(2-sulfanylethyl) amino]-2-(polystyrylmethyl)-3(*2H*)-pyridazinone **13** in 5 cm^3 of anhydrous DMF and add 0.80 cm^3 = 10.8 mmol trifluoroacetic acid (TFA) while stirring. Measure in 0.5 g = 1.32 mmol HTMAB and stir the reaction mixture for 3 hours at 80°C. Cool the reaction mixture to room temperature and filter using a G3 glass suction filter. Wash the polymer with 5 cm^3 of water, 2 × 3 cm^3 of DMF, 5 cm^3 of water, 5 cm^3 of methanol, 5 cm^3 of a methanol:toluene 1:1 mixture, 5 cm^3 of methanol, 5 cm^3 of a methanol:toluene 1:1 mixture, 2 × 5 cm^3 of methanol, and 2 × 5 cm^3 of diethyl ether. Dry the polymer in a drying apparatus at 50°C until you receive a constant mass.

Yield
0.53 g of polymer **21** (yellow in color)

Infrared spectrum of the cleaved polymer 21
νO–H, 3416 cm^{-1}; ν_{arom}C–H, 3020 cm^{-1}; ν_{aliph}C–H, 2922 cm^{-1}; νC=O, 1622 cm^{-1}; ν_{arom}C=C, 1512 and 1450 cm^{-1}; γ_{arom}C–H, 812 and 704 cm^{-1}

IDS (product – starting material)
Appearing bands: νC–H, 3434 cm^{-1}; νC=O, 1678 cm^{-1}; νC–C, 1196 and 1132 cm^{-1}
Disappearing bands: δN–H, 1560 and 1542 cm^{-1}
The combined filtrate was evaporated under reduced pressure.

Yield
0.2 g of a yellow oil (100%) (R_f (diethyl ether): 0.83. Elution by acetone was used during column chromatography to get the main product.

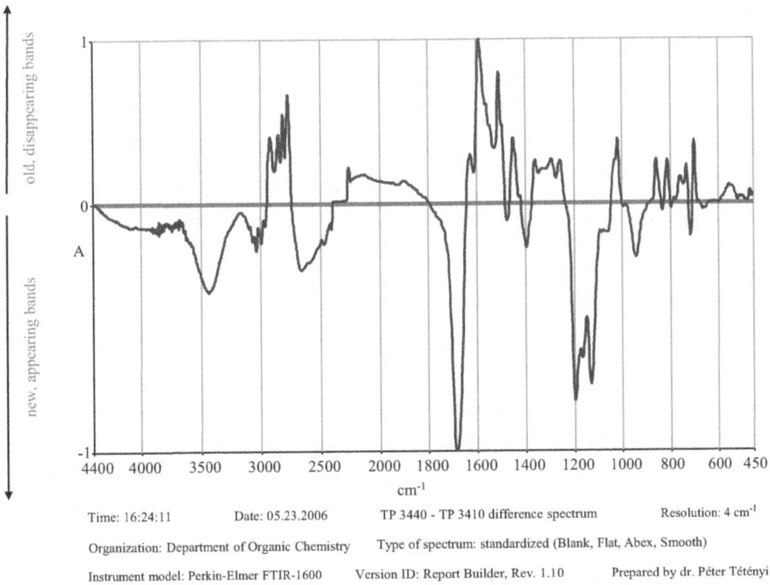

Infrared spectrum of the first cleaved product, product 22
νO–H, 3426 cm^{-1}; ν_{aliph}C–H, 2960, 2936, and 2874 cm^{-1}; νC=O, (ester) 1736 cm^{-1}; νC=O(amide), 1656 cm^{-1}; γ_{arom}C–H, 836 cm^{-1}

Infrared spectrum of the second cleaved product, product 23
νO–H, 3434 cm^{-1}; ν_{aliph}C–H, 2920 and 2850 cm^{-1}; νC=O(amide), 1690 cm^{-1}; νC–O, 1202 cm^{-1}; γ_{arom}C–H, 842, 802, and 724 cm^{-1}

^{1}H NMR spectrum (chloroform-d [$CDCl_3$]) of product 23
3.37 δ ppm s 1 H, 3.05 δ ppm s 1 H, 1.30 δ ppm s 2 H, 0.90 δ ppm m 8 H

2.14 Preparation of 4-Chloro-5-[(2-sulfanylethyl)amino]-3(2*H*)-pyridazinone 23 by a Strong Base (Cleavage Reaction)

Cleavage from the polymer: Equip a one-necked, round-bottomed flask of 25 cm^3 with a Liebig condenser. Suspend 0.41 g = 0.5 mmol 4-chloro-5-[(2-sulfanylethyl)amino]-2-(polystyrylmethyl)-3(*2H*)-pyridazinone **13** in 10 cm^3 of a methanol:toluene 1:1 mixture and add 11 cm^3 = 11 mmol sodium ethanolate 1 M solution in ethanol while stirring. Boil the reaction mixture for 3 hours at 100°C. Cool the reaction mixture to room temperature and filter using a G3 glass suction filter. Wash the polymer with 5 cm^3 of water, 5 cm^3 of methanol, 5 cm^3 of a methanol:toluene 1:1 mixture, 5 cm^3 of methanol, 5 cm^3 of a methanol:toluene 1:1 mixture, 2 × 5 cm^3 of methanol, and 2 × 5 cm^3 of diethyl ether. Dry the polymer in a drying apparatus at 50°C until you receive a constant mass.

Yield

0.34 g of a yellowish-brown polymer

The combined filtrate was evaporated under reduced pressure to give 2.39 g yellow crystals. It was extracted by a butanol:ethyl acetate 2:1 mixture, then by water, dried over sodium sulfate, and filtered and evaporated, yielding 0.05 g of a yellow oil (39%) (R_f (diethyl ether): 0.84. Elution by diethyl ether was used during column chromatography to get the main products. Two products were isolated: the first cleaved product was 0.023 g (22.4%), with R_f = 0.92; while the second cleaved product was 0.025 g (13.1%), with R_f = 0.84.

Infrared spectrum of the cleaved polymer

νO–H, 3420 cm^{-1}; ν_{arom}C–H, 3021 cm^{-1}; ν_{aliph}C–H, 2923 and 2860 cm^{-1}; νC=O, 1596 cm^{-1}; ν_{arom}C=C, 1508 and 1452 cm^{-1}; γ_{arom}C–H, 813, 761, and 701 cm^{-1}

Infrared spectrum of the first cleaved product, product 14

ν_{aliph}C–H, 2958, 2927, and 2858 cm^{-1}; νC=O, (ester) 1729 cm^{-1}; νC–O, 1274 cm^{-1}

^{1}H NMR spectrum ($CDCl_3$) of product 14

7.67 δ ppm dd 1 H, 7.54 δ ppm dd 1 H, 4.22 δ ppm qd 2 H, 1.69 δ ppm m 2 H, 1.41 δ ppm m 11 H, 0.90 δ ppm m 6 H

Infrared spectrum of the second cleaved product

ν_{aliph}C–H, 2954, 2927, and 2856 cm^{-1}; νC=O (ester), 1728 cm^{-1}; νC=O(amide), 1653 cm^{-1}; νC–O, 1288 cm^{-1}

^{1}H NMR spectrum ($CDCl_3$) of the second cleaved product

7.80 δ ppm s 1 H, 7.69 δ ppm m 1 H, 7.53 δ ppm m 1 H, 7.36 δ ppm m 1 H, 7.27 δ ppm m 1 H, 4.24 δ ppm m 2 H, 3.81 δ ppm s 2 H, 3.77 δ ppm d 1 H; 1.39 δ ppm m 6 H, 1.31 δ ppm m 20 H, 0.88 δ ppm m 9 H

2.15 Preparation of 4-Chloro-3(2*H*)-pyridazinone 26 (Cleavage Reaction)

Equip a one-necked, round-bottomed flask of 25 cm^3 with a Liebig condenser. Suspend 2.1 g = 3.75 mmol 4-chloro-2-(4-polystyryl(hydroxymethyl)phenyloxyethyl)-3(2*H*)-pyridazinone **19** in 10 cm^3 of anhydrous DMF and add 0.65 g = 1.72 mmol HTMAB and 1.11 cm^3 = 15 mmol TFA while stirring. Stir the reaction mixture for 3 hours at 80°C. Cool the reaction mixture to room temperature and filter using a G3 glass suction filter. Wash the polymer with 5 cm^3 of water, 2 × 3 cm^3 of DMF, 5 cm^3 of water, 5 cm^3 of methanol, 5 cm^3 of a methanol:toluene 1:1 mixture, 5 cm^3 of methanol, 5 cm^3 of a methanol:toluene 1:1 mixture, 2 × 5 cm^3 of methanol, and 2 × 5 cm^3 of diethyl ether. Dry the polymer in a drying apparatus at 50°C until you receive a constant mass.

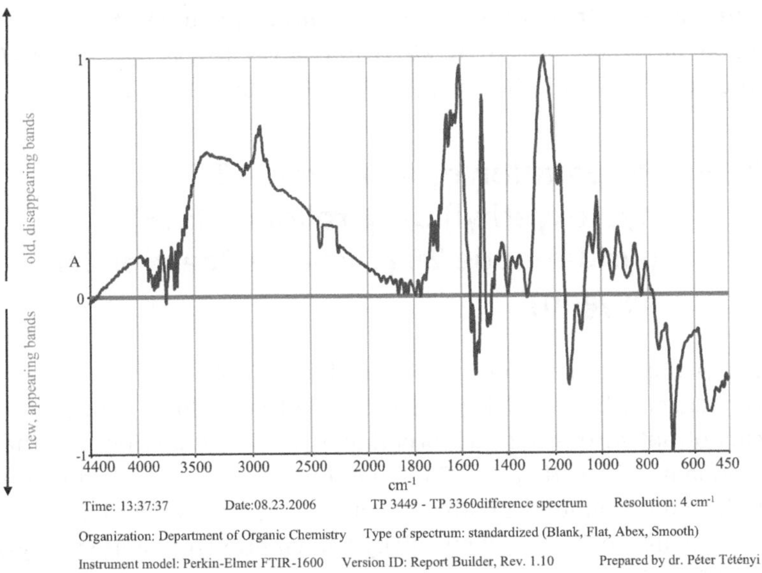

Yield

1.73 g of polymer **25**, brown in color, (yield 85.3% according to infrared analysis) with loading of 1.63 mmol/g

Infrared spectrum of the cleaved polymer 25
νO–H, 3386 cm^{-1}; ν_{arom}C–H, 3020 cm^{-1}; ν_{aliph}C–H, 2918 and 2852 cm^{-1}; νC=O, 1600 cm^{-1}; ν_{arom}C=C, 1502 and 1448 cm^{-1}; γ_{arom}C–H, 820, 758, and 698 cm^{-1}

IDS (product – starting material)
Appearing bands: $\nu_{arom.}$C–H, 3044 cm^{-1}; ν_{aliph}C–H, 2926 cm^{-1}; δN–H, 1540 cm^{-1}; νC–C, 1142 cm^{-1}; γ_{arom}C–H, 754 and 694 cm^{-1}
Disappearing bands: νC=O, 1660, 1640, and 1608 cm^{-1}; νC–C, 1510 cm^{-1}; νC–O, 1250 cm^{-1}

The combined filtrate was evaporated under reduced pressure. It was then extracted by chloroform two times and then by a butanol:toluene 1:1 mixture two times. Then the combined organic phase was extracted by water, dried over anhydrous sodium sulfate, filtered, and evaporated. The raw product was column-chromatographed twice, resulting in 0.38 g (76.5%) of a yellow, quickly crystallizing oil.

Infrared spectrum of the cleaved product, product 26
νO–H, 3436 cm^{-1}; ν_{aliph}C–H, 2926 and 2856 cm^{-1}; $\nu_{saturated\ amide}$C=O, 1684 cm^{-1}; νC–O, 1206 cm^{-1}; γ_{arom}C–H, 844, 804, and 724 cm^{-1}

2.16 Preparation of 4-Chloro-2-(2-hydroxyethyl)-5-pyrrolidino-3(2*H*)-pyridazinone 28 (Cleavage Reaction by Oxygen)

Equip a one-necked, round-bottomed flask of 25 cm^3 with a Liebig condenser. Dissolve 3.3 g = 13.05 mmol crystalline iodine in 12 cm^3 of anhydrous 1,2-dichloroethane (DKE) by stirring and then gradually add 0.48 g = 13.05 mmol sodium tetrahydridoborate to it. After 5 minutes of stirring, add 0.74 g = 0.76 mmol 4-chloro-5-pyrrolidino-2-(4-polystyryl(hydroxymethyl)phenyloxyethyl)-3(2*H*)-pyridazinone **12** to it. Stir the reaction mixture for 3.5 hours at 100°C. Cool the reaction mixture to room temperature and filter using a G3 glass suction filter. Wash the polymer with 5 cm^3 of water, 5 cm^3 of methanol, 5 cm^3 of a methanol:toluene 1:1 mixture, 5 cm^3 of methanol, 5 cm^3 of a methanol:toluene 1:1 mixture, 2 × 5 cm^3 of methanol, and 2 × 5 cm^3 of diethyl ether. Dry the polymer in a drying apparatus at 50°C until you receive a constant mass.

Yield

0.82 g of polymer **27**, brown in color, (yield 72% according to infrared analysis) with loading of 0.97 mmol/g

Infrared spectrum of polymer 27

νO–H, 3422 cm^{-1}; ν_{arom}C–H, 3022 cm^{-1}; ν_{aliph}C–H, 2920 cm^{-1}; $\nu_{arom\ twin\text{-}ionic}$C=O, 1616 cm^{-1}; ν_{arom}C=C, 1508 and 1452 cm^{-1}; νC–O, 1238 cm^{-1}; γ_{arom}C–H, 820, 758, and 698 cm^{-1}

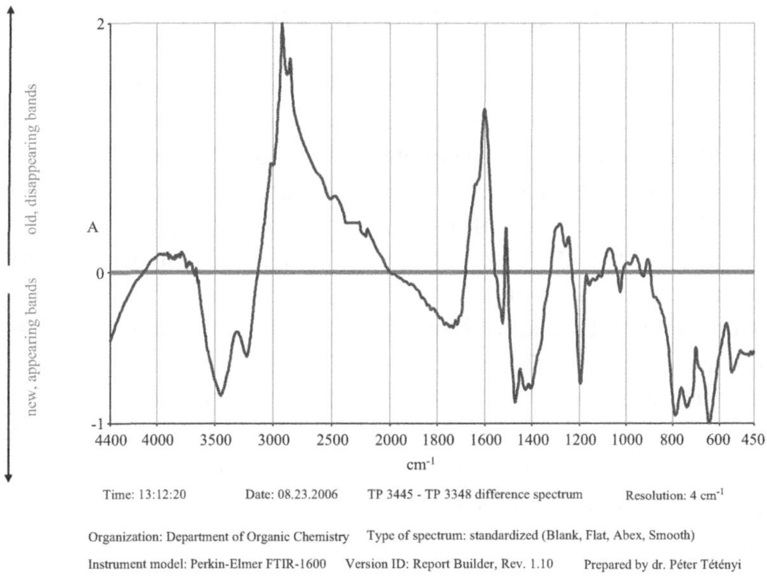

IDS (product – starting material)

Appearing bands: νO–H, 3446 cm^{-1}; νN–H, 3224 cm^{-1}; ν_{arom}C–C, 1470 and 1426 cm^{-1}; νC–O, 1194 cm^{-1}; γ_{arom}C–H, 786 and 738 cm^{-1}

Disappearing bands: ν_{aliph}C–H, 2920 and 2850 cm^{-1}; ν_{arom}C–C, 1600 cm^{-1}

The mother liquor was decolorized by the addition of solid sodium thiosulfate and then was evaporated under reduced pressure. The distillation residue was extracted by chloroform two times and then by a chloroform:butanol 1:1 mixture two times. Then the combined organic phase was extracted by water, dried over anhydrous sodium sulfate, filtered, and evaporated, yielding 0.04 g (13.2%) of a yellow, quickly crystallizing oil.

Infrared spectrum of the cleaved product, product 28
νO–H, 3448 cm^{-1}; ν_{aliph}C–H, 2924 and 2854 cm^{-1}; ν_{amide}C=O, 1688 cm^{-1}; ν_{arom}C=C, 1464 cm^{-1}; νC–O, 1202 cm^{-1}; γ_{arom}C–H, 838, 802, and 724 cm^{-1}

2.17 Preparation of 4-Chloro-2-(2-hydroxyethyl)-3(2H)-pyridazinone 29 (Cleavage Reaction by Oxygen)

Equip a one-necked, round-bottomed flask of 25 cm^3 with a Liebig condenser. Dissolve 4.17 g = 16.5 mmol crystalline iodine in 15 cm^3 of anhydrous DKE by stirring and then gradually add 0.61 g = 16.5 mmol sodium tetrahydridoborate to it. After 5 minutes of stirring, add 1.15 g = 0.96 mmol 4-chloro-2-(4-polystyryl(hydroxymethyl) phenyloxyethyl)-3(*2H*)-pyridazinone **19** to it. Stir the reaction mixture for 3.5 hours at 100°C. Cool the reaction mixture to room temperature and filter using a G3 glass suction filter. Wash the polymer with 5 cm^3 of water, 5 cm^3 of methanol, 5 cm^3 of a methanol:toluene 1:1 mixture, 5 cm^3 of methanol, 5 cm^3 of a methanol:toluene 1:1 mixture, 2 × 5 cm^3 of methanol, and 2 × 5 cm^3 of diethyl ether. Dry the polymer in a drying apparatus at 50°C until you receive a constant mass.

Yield
1.09 g of polymer **27**, light brown in color, (yield 78.2% according to infrared analysis) with loading of 0.77 mmol/g

Infrared spectrum of polymer 27
νO–H, 3421 cm^{-1}; ν_{arom}C–H, 3022 cm^{-1}; ν_{aliph}C–H, 2918 cm^{-1}; $\nu_{arom\ twin\text{-}oinic}$C=O, 1598 cm^{-1}; ν_{arom}C=C, 1489 and 1450 cm^{-1}; νC–O, 1232 cm^{-1}; γ_{arom}C–H, 823, 758, and 700 cm^{-1}

IDS (product – starting material)
Appearing bands: –
Disappearing bands: νC=O, 1686 cm^{-1}; νC–O, 1208 cm^{-1}; νC–C, 1138 cm^{-1}

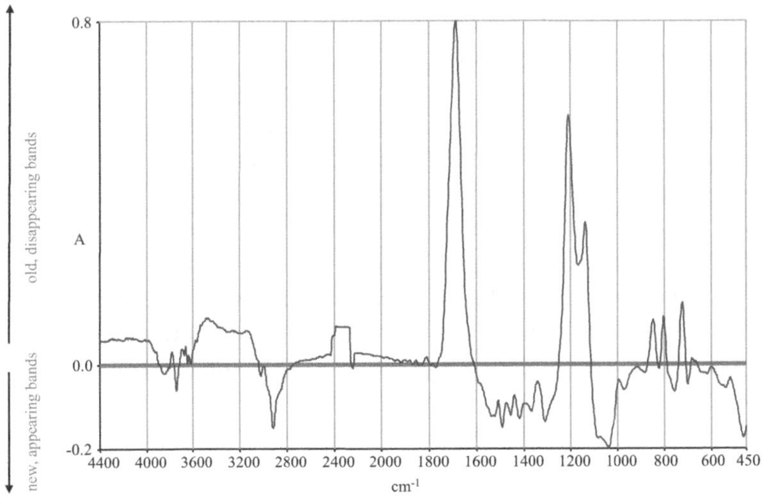

Time: 18:08:57 Date: 10.14.2006 TP 3453/A - TP 3451/A difference spectrum Resolution: 4 cm^{-1}

Organization: Department of Organic Chemistry Type of spectrum: standardized (Blank, Flat, Abex, Smooth)

Instrument model: Perkin-Elmer FTIR-1600 Version ID: Report Builder, Rev. 1.10 Prepared by dr. Péter Tétényi

The mother liquor was decolorized by the addition of solid sodium thiosulfate and then was evaporated under reduced pressure. The distillation residue was extracted by chloroform two times and then by a chloroform:butanol 1:1 mixture two times. Then the combined organic phase was extracted by water, dried over anhydrous sodium sulfate, filtered, and evaporated, yielding 0.1 g (60%) of a yellow, quickly crystallizing oil. This was column-chromatographed, eluted by dichloromethane, recrystallized from diethyl ether, and filtered, resulted in 0.05 g (30%) of yellow crystals **29**.

Infrared spectrum of the cleaved product, product 29
νO–H, 3422 cm^{-1}; ν_{aliph}C–H, 2958, 2926, and 2856 cm^{-1}; ν_{amide}C=O, 1734 and 1650 cm^{-1}; νC–O, 1260 cm^{-1}; νC–C, 1096 and 1026 cm^{-1}; γ_{arom}C–H, 804 cm^{-1}

2.18 Preparation of 4-Chloro-2-(2-hydroxyethyl)-3(2*H*)-pyridazinone 29 (Cleavage Reaction by Oxygen; Checking by ^{1}H NMR)

Equip a one-necked, round-bottomed flask of 25 cm^3 with a Liebig condenser. Dissolve 3.8 g = 15 mmol crystalline iodine in 20 cm^3

anhydrous DKE by stirring and then gradually add 0.55 g = 15 mmol sodium tetrahydridoborate to it. After 5 minutes of stirring, add 0.84 g = 1.1 mmol 4-chloro-2-(4-polystyryl(hydroxymethyl) phenyloxyethyl)-3(*2H*)-pyridazinone **19** to it. Stir the reaction mixture for 3 hours at 90°C. Cool the reaction mixture to room temperature and then pour it into a mixture of 10 cm^3 of water and 10 g of ice. Filter the reaction mixture through a G3 glass suction filter. Wash the polymer with 5 cm^3 of water, 5 cm^3 of methanol, 5 cm^3 of a methanol: toluene 1:1 mixture, 5 cm^3 of methanol, 5 cm^3 of a methanol: toluene 1:1 mixture, 2 × 5 cm^3 of methanol, and 2 × 5 cm^3 of diethyl ether. Dry the polymer in a drying apparatus at 50°C until you receive a constant mass.

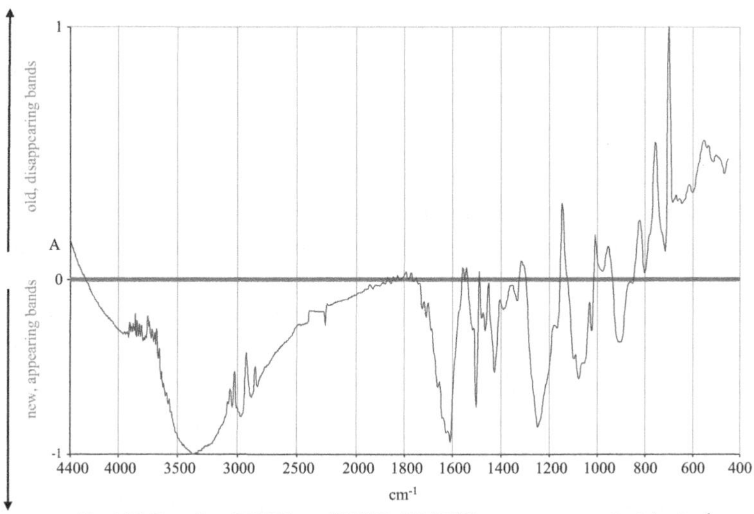

Time: 14:22:58. Date: 10.27.2006. TP 3454/A - TP 3409 difference spectrum Resolution: 4 cm^{-1}

Organization: Department of Organic Chemistry Type of spectrum: standardized (Blank, Flat, Abex, Smooth)

Instrument model: Perkin-Elmer FTIR-1600 Version ID: Report Builder, Rev. 1.10 Prepared by dr. Péter Tétényi

Yield

0.85 g of polymer **27**, light brown in color, (yield approximately 48%) with loading of 0.7 mmol/g

The mother liquor was decolorized by the addition of solid sodium hydrogen sulfate and then was evaporated under reduced pressure. The distillation residue was extracted by a chloroform: butanol 3:1 mixture three times. Then the combined organic phase was extracted by water, dried over anhydrous sodium sulfate, filtered, and evaporated, yielding 2.1 g of a black oily crystalline mass. This was column-chromatographed twice and eluted by diethyl ether and then by a 1:1 mixture of ACN and methanol. Two fractions were isolated. One was 0.04 g (13.5%) of brown crystals **29**, with $R_f = 0.22$ in a diethyl ether:methanol mixture of 10 to 1; the ^{1}H NMR spectrum had the following chemical shifts: (methanol-d4 [MeOD]): 8.20 δ ppm dd 2 H, 8 δ ppm dd 1 H, 3.15 δ ppm m 2 H, 3.05 δ ppm br s 3 H, 1.96 δ ppm d 2 H. The other fraction was 0.103 g (35%) of light-brown crystals, with $R_f = 0.42$ in a diethyl ether:methanol mixture of 10 to 1. The ^{1}H NMR spectrum had the following chemical shifts: (MeOD): 4.65 δ ppm s 2 H, 1.28 δ ppm s 2 H, 0.87 δ ppm t 1 H. The structure is unidentified.

2.19 Preparation of 4-Chloro-2-(2-hydroxyethyl)-5-[(2-sulfanylethyl)amino]-3(2*H*)-pyridazinone 30 (Cleavage Reaction by Oxygen)

Equip a one-necked, round-bottomed flask of 25 cm^3 with a Liebig condenser. Dissolve 3.2 g = 12.7 mmol crystalline iodine in 11.5 cm^3 anhydrous DKE by stirring and then gradually add 0.47 g = 12.7 mmol sodium tetrahydridoborate to it. After 5 minutes of stirring, add 1.21 g = 0.74 mmol 4-chloro-5-[(2-sulfanylethyl)amino]-2-(4-polystyryl(hydroxymethyl)phenyloxyethyl)-3(2*H*)-pyridazinone **15** to it. Stir the reaction mixture for 3 hours at 100°C. Cool the reaction mixture to room temperature and filter using a G3 glass suction filter. Wash the polymer with 5 cm^3 of water, 5 cm^3 of methanol, 5 cm^3 of a methanol:toluene 1:1 mixture, 5 cm^3 of methanol, 5 cm^3 of a methanol:toluene 1:1 mixture, 2 × 5 cm^3 of methanol, and 2 × 5 cm^3 of diethyl ether. Dry the polymer in a drying apparatus at 50°C until you receive a constant mass.

Yield

1.16 g of polymer **27**, light brown in color, (yield 49.1% according to infrared analysis) with loading of 0.65 mmol/g

Infrared spectrum of the cleaved polymer 27

νO–H, 3390 cm^{-1}; ν_{arom}C–H, 3020 cm^{-1}; ν_{aliph}C–H, 2918 cm^{-1}; νC=O, 1604 cm^{-1}; ν_{arom}C=C, 1502 and 1448 cm^{-1}; νC–O, 1236 cm^{-1}; γ_{arom}C–H, 820, 760, and 700 cm^{-1}

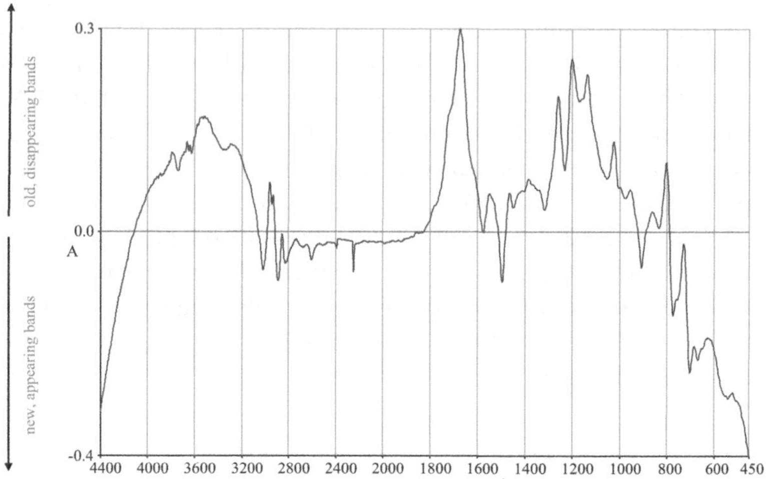

Time: 16:21:33 Date:11.18. 2006. TP 3455/A - TP 3450/A difference spectrum Resolution: 4 cm^{-1}

Organization: Department of Organic Chemistry Type of spectrum: standardized (Blank, Flat, Abex, Smooth)

Instrument model: Perkin-Elmer FTIR-1600 Version ID: Report Builder, Rev. 1.10 Prepared by dr. Péter Tétényi

IDS (product – starting material)

Appearing bands: νN–H and νO–H, 3368 cm^{-1} (overlapping); ν_{arom}C–H, 3040 cm^{-1}; ν_{aliph}C–H, 2970, 2882, and 2828 cm^{-1}; νC=O, 1660 and 1610 cm^{-1}; νC=C, 1502 cm^{-1}; νC–O, 1248 cm^{-1}; γ_{arom}C–H, 700 cm^{-1}

Disappearing bands: νC–C, 1146 cm^{-1}; γ_{arom}C–H, 758 and 700 cm^{-1}

The combined filtrate was evaporated under reduced pressure and then was extracted by chloroform two times and then by a butanol: toluene 1:1 mixture two times. Then the combined organic phase was extracted by water, dried over anhydrous sodium sulfate, filtered, and evaporated. The raw product was column-chromatographed twice, resulting in 0.08 g (43.3%) of a yellow, quickly crystallizing oil **30** and 0.045 g (26.3%) of a brown oil by-product.

Infrared spectrum of the cleaved product, product 30

νO–H, 3416 cm^{-1}; ν_{aliph}C–H, 2924 and 2856 cm^{-1}; ν_{ester}C=O, 1738 cm^{-1}; $\nu_{lactame}$C=O, 1640 cm^{-1}; νC–O, 1208 cm^{-1}; γ_{arom}C–H, 806 cm^{-1}

2.20 Preparation of 4-Chloro-5-iodo-2-(2-hydroxyethyl)-3(2*H*)-pyridazinone 31 (Cleavage Reaction by Oxygen; Checking by ^{1}H NMR)

Equip a one-necked, round-bottomed flask of 50 cm^3 with a Liebig condenser. Dissolve 7.95 g = 31.4 mmol crystalline iodine in 42 cm^3 of anhydrous DKE by stirring and then gradually add 1.13 g = 31.4 mmol sodium tetrahydridoborate to it. After 5 minutes of stirring, add 2 g = 1.32 mmol 4-chloro-5-iodo-2-(4-polystyryl(hydroxymethyl) phenyloxyethyl)-3(2*H*)-pyridazinone **10** to it. Stir the reaction mixture for 2 hours at 110°C. Cool the reaction mixture to room temperature and then pour it into a mixture of 10 cm^3 of water and 10 g of ice. Filter the reaction mixture through a G3 glass suction filter. Wash the polymer with 5 cm^3 of water, 5 cm^3 of methanol, 5 cm^3 of a methanol:toluene 1:1 mixture, 5 cm^3 of methanol, 5 cm^3 of a methanol:toluene 1:1 mixture, 2 × 5 cm^3 of methanol, and 2 × 5 cm^3 of diethyl ether. Dry the polymer in a drying apparatus at 50°C until you receive a constant mass.

Yield

1.95 g of polymer **27**, light brown in color, (yield approximately 32%) with loading of 0.64 mmol/g

The mother liquor was decolorized by the addition of solid sodium hydrogensulfate and then was evaporated under reduced pressure. The distillation residue was extracted by a chloroform:butanol 3:1 mixture three times. Then the combined organic phase was extracted by water, dried over anhydrous sodium sulfate, filtered, and evaporated, yielding 0.1 g of a brown oil. This was column-chromatographed twice and eluted by a diethyl ether:chloroform 3:1 mixture, yielding 0.173 g (32%) of yellow crystals **31**, with R_f = 0.81 in a diethyl ether:methanol mixture of 10 to 1. The ^{1}H NMR spectrum had the following chemical shifts: ($CDCl_3$): 7.16 δ ppm m 2 H, 3.05 δ ppm br s 2 H, 2.31 δ ppm d 1 H, 1.75 δ ppm s 3 H, 1.32 δ ppm s 19 H, 0.93 δ ppm d 8 H.

Chapter 3

Results

In the following, a conventional solid-phase synthetic strategy using polystyrene beads cross-linked with 1% divinylbenzene was going to be built. Although chloromethyl polystyrene (bought from FLUKA) was used in a few model experiments, usually we functionalized the polystyrene support by a S_EAr reaction to provide loading sufficiently high for preparative purposes. Thus, approximately every second or third ring of the polystyrene had a linker (the loading was 48.5% ring substitution for our chloromethyl polystyrene while 31.3% ring substitution for the polystyrene with the phenol ether linker, according to the elemental analysis results; such polymers were not sold by any company).

On the basis of our previous research both on soluble heterocyclic chemistry and on the solid-phase syntheses of linear/branched chains, the following strategy was established (see Scheme 3.1).

This was the simplest scheme to be applied; however, additional steps with polymer-supported substrates were better suited to our needs.

Selection of the substrate was a complicated problem, since we were aware of the easy double reaction with polymer-supported substrates. The higher the loading, the greater the chances of making new intrapolymer cross-links if bifunctional reagents are used. Unfortunately, the simplest ring closure reactions of pyridazines are reactions of (usually symmetric) bifunctional reagents. Under

Polymers and Pyridazines
Péter Tétényi

ISBN 978-981-4800-47-1 (Paperback), 978-0-429-29646-8 (eBook)
www.jennystanford.com

preparative conditions (if loading of the starting polymer was at least 1 mmol/g), the undesired double reaction with bifunctional reagents[a] looked unavoidable (see also Ref. [11]). Thus, a pyridazine ring was prepared by a soluble reaction according to reaction Eq. 3.1 [15] and then this ring was used for attachment reactions:

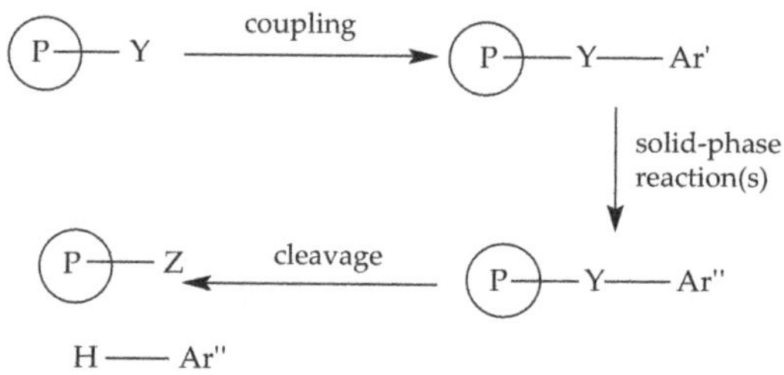

Here, (P) was a 1% cross-linked polystyrene in bead form, Y was the linker, Ar′ was the proposed substrate with a pyidazine ring, Ar″ was the supported and modified substrate with a pyridazine ring, and Z was the modified linker generated in the cleavage step.

Scheme 3.1 General sequence for the solid-phase synthesis of pyridazine derivatives (based on Merrifield's method).[b]

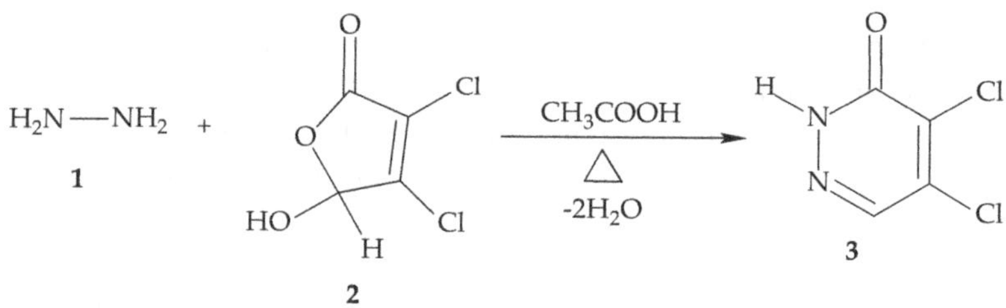

Reaction Equation 3.1 Soluble synthesis of 4,5-dichloro-3(*2H*)-pyridazinone **3**.

3.1 Swelling Experiments

We soon realized that swelling played an important role in the chemical reactions of solid-phase synthesis. Swelling must be used

[a]That is, the double reaction of any of the reagent molecules with two separate polymer-supported chains, resulting in the formation of a (larger) ring inside the polymer beads. It would be also advantageous from the viewpoint of the enthropy changes.

[b]This was only a theoretical, general scheme. Thus, the compounds were not numbered.

to explain chemical properties of various polymers throughout this book [16]. Thus, it became inevitable to carry out additional swelling measurements with the new polymers, because swelling of the support had a strong influence on the reaction rates. Few methods for swelling measurements can be found in the literature. However, usually these nondynamic measurements [17–19] are carried out at room temperature for 12–48 hours in a single-solvent system. We measured swelling of the functionalized supports in a pipette closed at the tip and always poured a suspension of the polymer in the given solvent into a pipette immersed in a test tube filled with the appropriate solvent of the proper temperature (not only at room temperature!). Swelling results were evaluated after about 10 minutes (dynamic swelling measurement); thus, these results could provide much more information on how the polymer would behave in the real conditions of the experiments.

This problem was a significant one, since complete precipitation of the functionalized polymers was observed in some steps of the experiments mentioned in this chapter and in these cases, the total unsuitable conditions must be mentioned, which are as follows:

- If we carry out preparation of polymers **5** or 7 without any dipolar aprotic solvent (e.g., without *N,N*-dimethylformamide [DMF]).
- If we carry out reaction of polymer **5** with an excess of hydrazine, without acetonitrile (ACN), precipitation takes place; salt formation would result in shrinkage without any dipolar aprotic solvent.
- Polymer **9** has a poor swelling ability in protic solvents (e.g., in butanol), and because its formation took place in ethylene glycol (EG), addition of a quaternary catalyst was needed for completion of the reaction (increasing temperature did not increase the swelling ability).
- Polymer **18** has an exceptionally poor swelling ability in dipolar aprotic solvents (e.g., in DMF); thus, it would have been better to prepare it in a toluene:DMF mixture (instead of using DMF alone).
- If we carry out the cleavage reaction by the N in an apolar solvent (e.g., in chloroform).

These bad conditions should be always avoided. Experimental conditions of swelling were given at the beginning of Chapter 2. Swelling results are summarized in Table 3.1 as well as graphically shown in Charts 3.1 and 3.2:

Table 3.1 Dynamic swelling measurements: swelling *x* times the volume measured in the dry stage

	Swelling in toluene		**Swelling in butanol**		**Swelling in DMF**	
Dipole moment	μ = 0.36 D*		μ = 1.66 D*		μ = 3.82 D*	
Polymer	At RT	At 120°C	At RT	At 120°C	At RT	At 120°C
4	3.1	2.3	1.8	2.3	1.5	2.2
5	1.8	1.9	2.6	2.6	1.4	1.7
9	1.5	1.6	0.66	0.56	1.3	1.7
11	1.8	1.9	2.1	2.4	1.9	2.4
13	1.9	2.6	2.1	2.1	2	2.2
16	2.2	2	2.2	2.3	1.4	2
18	1.6	1.9	1.7	1.9	1.2	1.2
6	2.3	2.5	1.5	1.8	3.6	3.3
7	2.1	2.2	3.1	3.8	2.8	2.9
10	5.2	5.3	14**	2.9**	10**	3.4**
12	3.3	4.3	3.1	3.2	1.8	1.8
15	2.6	3.7	11.2***	2.7	5	3.6
17	2.7	2.9	11.2***	3	11.2***	4.4
19	2.3	2.9	3.6	2.6	3.6	3.1

*Dipole moments of the solvents were taken from *CRC Handbook of Chemistry and Physics*, 67th Ed. Only values measured in gas phase were available.
**With only 0.1 g of the polymer sample.
***These polymers (0.2 g each) were homogeneously distributed in the whole volume (2 cm^3) of the pipette in the given solvents. Thus, no interphase could be observed even after 30 minutes. In these cases, swelling was considered to be 11.2 times (an excellent but unexpected value).

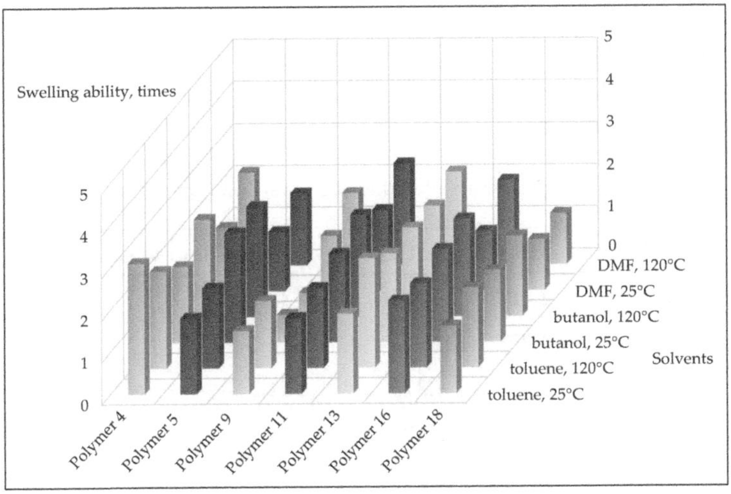

Chart 3.1 Swelling ability of polymers derived from chloromethylpolystyrene.

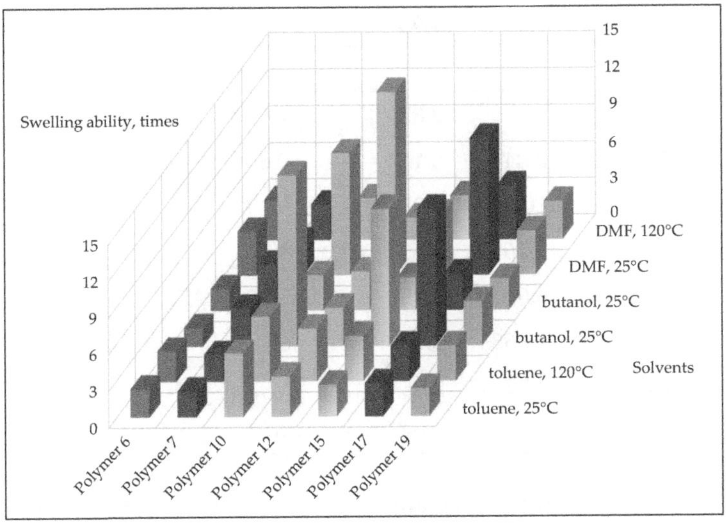

Chart 3.2 Swelling ability of polymers derived from a polymer with a multidetachable linker.

However, some general conclusions could be drawn from the swelling results of Table 3.1. Moreover, an additional table (Table 3.2) was prepared from the measured swelling results of Table 3.1, taking into consideration the dipole moments of the solvents. Thus, the swelling number *S* was introduced; *S* was the average of the swelling values of the polymer at the given temperature, weighted by the dipole moments of the solvent(s). Swelling numbers were also dimensionless.

Table 3.2 Swelling numbers of various polymers (in two parts)

Polymer	*S* at RT	*S* at 120°C	Δ% gradient with signs
4	1.68	2.23	+32.7
5	1.77	1.97	+11.1
9	1.13	1.37	+21.2
11	1.95	2.37	+21.5
13	2.02	2.20	+8.9
16	1.68	2.09	+24.4
18	1.37	1.44	+5.1
average:	1.66	1.95	+17.6
6	3.4	2.82	-17
7	2.84	3.11	+9.6
10	10.84	3.38	-68.9
12	2.26	2.35	+4
15	6.61	3.35	-50.6
17	10.65	3.73	-65
19	3.52	2.95	-17.3
average:	5.73	3.09	-46

A significant difference appeared between the average swelling numbers of the polymers generated from polymer **4** (1.95 times at 120°C) and of the polymers generated from polymer **6** (3.09 times at 120°C). The swelling numbers at room temperature would not be evaluated in themselves since the reactions were always done by heating. Moreover, while the swelling of the polymers generated from polymer **4** increased by more than 17% on increasing the temperature from room temperature to 120°C, swelling of the parallel series of the polymers generated from polymer **6** showed an opposite tendency: it decreased by 46% on increasing the temperature. Swelling gradients of various polymers in the function of the temperature are graphically shown in Chart 3.3. This kind of swelling and swelling gradient differences characterized the linkers since the same support, the same substrates, and the same swelling solvents were used. Thus, it was proven from the point of view of swelling that it was worthwhile to introduce this new linker with phenol ether: it provided an average swelling number greater by 58.5% at 120°C, compared to the polymers generated from polymer **4**.

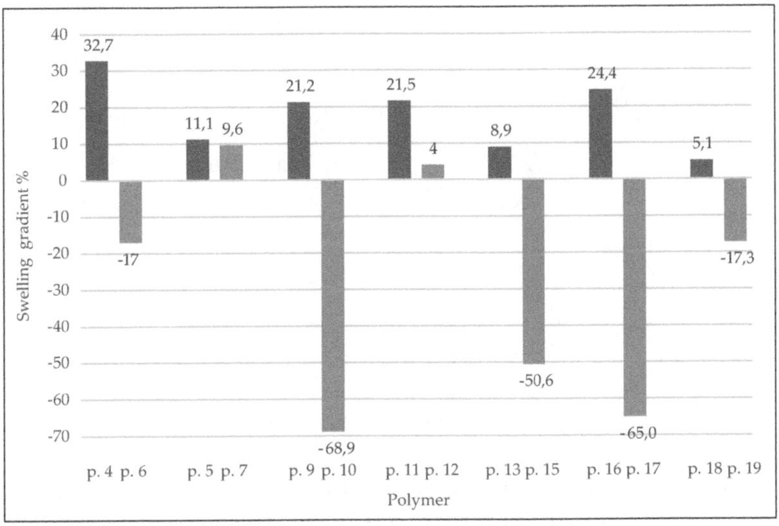

Chart 3.3 Relative swelling gradients of polymers; the effect of temperature.

Consequences were discussed of the greater swelling values for the given reactions.

3.2 Attachment Reactions

The following attachments, which were stable during the whole solid-phase synthesis, were going to be introduced. The alternatives are shown in Reaction Eqs. 3.2 and 3.3:

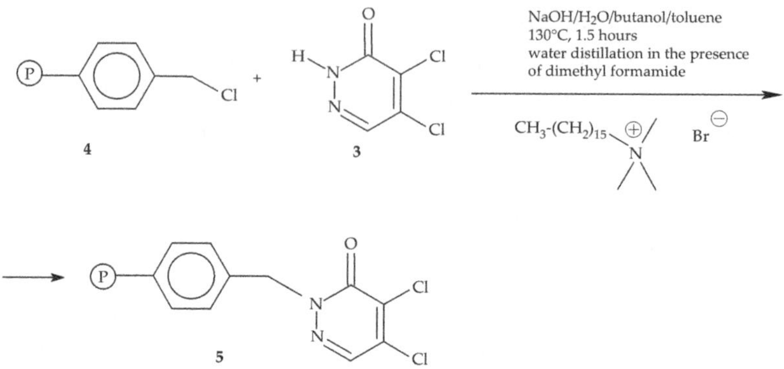

Reaction Equation 3.2 Attachment to chloromethyl polystyrene.

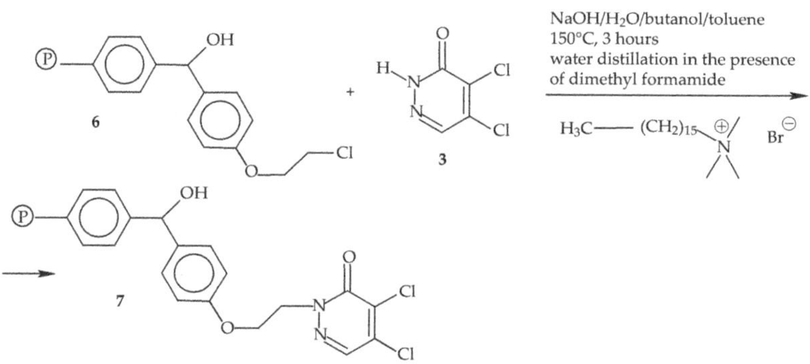

Reaction Equation 3.3 Attachment to the polymer by a phenol ether linker.

Both supported substrates (**5**, **7**) remained stable during most nucleophilic substitution reactions of the substrate. The synthetic processes needed for the preparation of polymer **6** and of the related polymer(P)-C_6H_4-SCH_2CH_2Cl (**8**) are not described in this book since referees considered these processes as known methods upon repeated attempts for publication. However, we could not a reach yield higher than 15% by using any traditional halogenation methods in polymer-supported synthesis; according to our halogenation method, the yield was usually around 70% or above [16] and it could be improved to 91% by proper repetition of the halogenation.

The constant parameters of the solvent system were water/alcohol/toluene, deprotonation by sodium hydroxide, and catalysis by 10% hexadecyltrimethylammonium bromide (HTMAB); and

DMF was added (see Reaction Eq. 3.2 and Table 3.3). Loading of the starting materials was checked by elemental analyses, by control reaction (pyridylation), and by Fourier-transform infrared spectroscopy [FTIR]. The changing factors are summarized in Table 3.3:

Table 3.3 Optimization experiments for the attachment to polymer **4**

A	B	C	D	E	F	G	H	I
1	1.06x	methanol	-	0.23	1	70	1	4
2	2.12x	methanol	-	0.23	1	70	18	45.2
3	2.12x	butanol	23.5	0.18 → 0.47	2	130	2.33	42.9
4	2.13x	butanol	23.5	0.18 → 0.31	2	140	1.5	59.5
5	2.13x	butanol	23.5	0.18 → 0.52	2	140	2	62.5
6	2.13x	butanol	11.5	0.46 → 0.36	2	140	3.5	100

A: Experiment number
B: Amount of sodium hydroxide to the reagent
C: Type of protic solvent
D: Ratio of DMF in %
E: Relative concentration of the reagent to the substrate
F: Reaction method: 1 boiling; 2 water distillation by a Dean–Stark trap
G: Temperature in °C
H: Reaction time in hours
I: Yield of attachment (see the text) in %

Yields were calculated from the standardized infrared spectra by various methods. Finally, yields for the attachment in Table 3.3 were determined either by manually calculating the lg A/A values of the carbonyl band of the nonstandardized spectra using the γC–H at 831 cm^{-1} as a reference band or by electronically calculating the area of the carbonyl band at 1638 cm^{-1} of the standardized spectra of the infrared difference spectrum (IDS) (product – starting material), setting it to zero value at the γ C–H reference band of 831 cm^{-1}. Maximum (100%) transformation was attributed to the relative carbonyl intensity or area (see experiment number 6, in Table 3.3) when the β CH_2Cl at A_{1264} disappeared completely.

The other calculation methods of yields did not prove to be reliable, often resulting in too-high values. Thus, the mass changes measured were useless; direct measurement of the areas of the chloromethyl band (β CH_2Cl at A_{1264}) resulted in too-high values if these were measured alone or in comparison to the γC–H reference band of 831 cm^{-1}. This was due to the fact that the β CH_2Cl band was a moderate one and, moreover, the baseline of the β CH_2Cl was never at 100% transmittance level even in the standardized spectra (more often, it was between 40% and 60%), resulting in an increased experimental error. Other reference bands could not be used, since most of the ν_{arom} C=C bands were strongly influenced by the presence of the substrate. On the other side, measuring the carbonyl intensity or area alone (without using any reference bands) was not reliable at all; results were the function of the amount of the polymer-supported substrate in the given IR tablet. Control reactions, like pyridylation, could not be used for checking course of the N-alkylation since the pyridylium band at 1648 cm^{-1} was a moderate band mostly overlapping with the new lactame band of the pyridazine substrate at 1653 cm^{-1} (increased error of the quantitative determination). It must be noted repeatedly that neither during the attachment reaction nor during any of the later steps did we observe any bands at 1382 and at 1139 cm^{-1} (false interpretation according to Ref. [3]). Courses of the N^2-alkylations by chloromethyl polystyrene are graphically depicted in Chart 3.4.

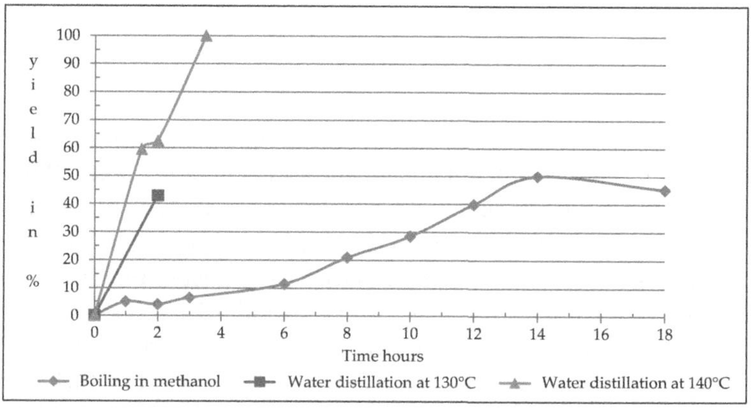

Chart 3.4 Course of N^2-alkylation for reactions with chloromethyl polystyrene 4.

Conditions of the experiment with water distillation at 130°C–140°C resulted in a reaction rate greater by 1 order than that of the experiment with simple boiling. The swelling number (*S*) of the polymer in the attachment reaction was slightly decreased (shrinkage happened from 2.23 to 1.97) and more DMF could not help swelling at all (see Table 3.1, polymer **5**). This precipitation happened despite the fact that reagent concentrations were much lower than substrate concentrations (0.18–0.52 times). But 10%–25% of DMF (column 3, Table 3.1) could help avoid precipitation of the inorganic compounds and salts (sodium hydroxide, sodium chloride, and HTMAB). These latter had poor solubility in toluene and butanol and thus—if DMF was omitted—could easily stop the attaching reaction by closing the main channels within the polymer beads.

Yields were the absolute numbers and the combined yields from (P)-C_6H_4-SCH_2CH_2OH or from (P)-C_6H_4-CHOH-C_6H_4-OCH_2-CH_2OH. Thus, an 80% yield meant a quantitative yield for the attachment reaction. Pyridylation could not be used here since there was not any spectroscopic change at 1650 cm^{-1} due to pyridylation of the sample built on a polymer with a multidetachable linker. Courses of the N^2-alkylations by the polymer with a multidetachable linker are graphically depicted in Chart 3.5.

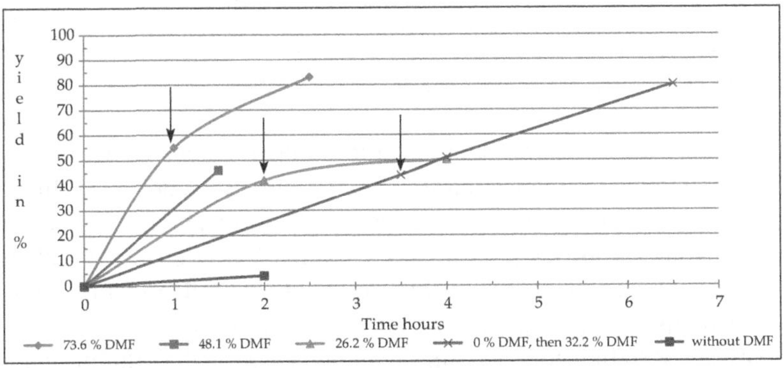

Chart 3.5 Course of N^2-alkylation for reactions with polymer **6** with a multidetachable linker.

The black arrows meant repetition of the reactions with the previously isolated polymer, using conditions similar to the ones before.

As stated before, the measured mass changes did not have any correlation to the actual yields. However, it did not mean that measuring the mass of the polymers was without any sense. Greater weights than 100% of the calculated mass indicated O-alkylation: in these cases, the product of the N-alkylation (like polymers **5** or **7**) was further alkylated on the lactame oxygen. If this extra mass increase was formed due to the O-alkylation, then it could be immediately removed by the attack of any nucleophile; polymers with only N-alkylation were much more stable against nucleophiles. The measured mass changes, together with the values of the polymer with the multidetachable linker, are depicted in Chart 3.6.

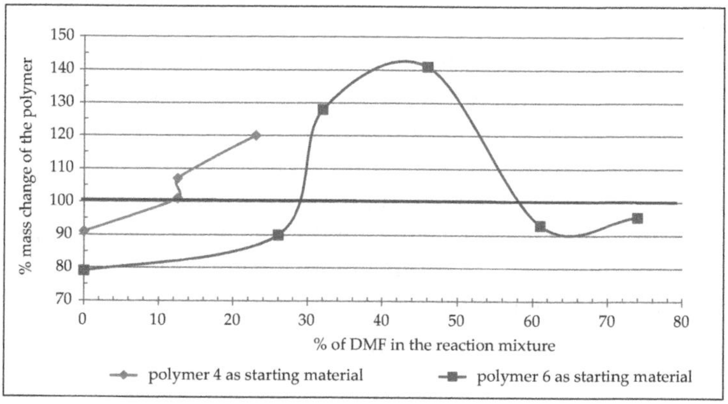

Chart 3.6 Double alkylation in the presence of DMF.

The constant parameters of the solvent system were water/butanol/toluene, deprotonation by sodium hydroxide 2.12 times the amount of the 4,5-dichloro-3(*2H*)-pyridazinone substrate, and catalyzation by HTMAB; water distillation was carried out by a Dean–Stark trap (see Reaction Eq. 3.3, Chart 3.5, and Table 3.5). The temperature was 150°C, and DMF was usually added during water distillation for the same reasons as we had in the previous attachment reaction. Usually more vigorous conditions were applied, and it proved to be considerably more difficult to attach the same substrate to a polymer with halogen in a nonactivated position (see Reaction Eq. 3.3).

Without DMF, the reaction was very slow and repetition of the attachment step was needed. What was quite strange with regard to the swelling experiments (Section 3.1) was that not only were the swelling numbers (*S*) greater for polymers **6** and **7** than for polymers **4** and **5**, respectively, but also in the hot reaction mixture, there was no shrinkage of the polymer beads, but on the contrary, their swelling state was improved during the reaction.

Loading of the starting materials was checked by elemental analyses and by infrared spectroscopy [FTIR]. Changing factors are summarized in Table 3.4; the same kind spectroscopic evaluation was applied for the reactions of polymers with multidetachable linkers as the one used for the preparation of polymer **5**.

Table 3.4 Optimization experiments for the attachment to polymer **6** with a multidetachable linker

A	B	C	D	E	F	G
7	-*	10x	0.18	130	2	4.5
8	73.6	1.8x	0.27	130	1	55
9	73.6**	1.68x	0.27	130	1.5	+27.8
10	48.1	3.36x	0.44	140	1.5	46.9
11	-	3.86x	0.51	140	3.5	44.2
12	32.2**	4.14x	0.21	140	3	+36.2
13	25.9	2.62x	0.13	150	2	42.5
14	26.5**	4.51x	0.24	150	2	+8.3

A: Experiment number
B: Ratio of DMF in %
C: Amount of dichloropyridazinone to the polymer
D: Initial relative concentration of the reagent to the substrate
E: Temperature in °C
F: Reaction time in hours
G: Yield of attachment (see the text) in %
*The linker had *S* instead of CH(OH)-C_6H_4-O (polymer **8**).
**Starting with the polymer isolated in the previous attachment reaction, previous row.

The following conclusions were drawn from Table 3.4. The presence of DMF was essential during the attachment reaction, and the higher the amount of DMF present, the higher was the reaction rate. However, we realized that once deprotonated, dichloropyridazinone **3** behaved as an ambident nucleophile (see Reaction Eq. 3.4).

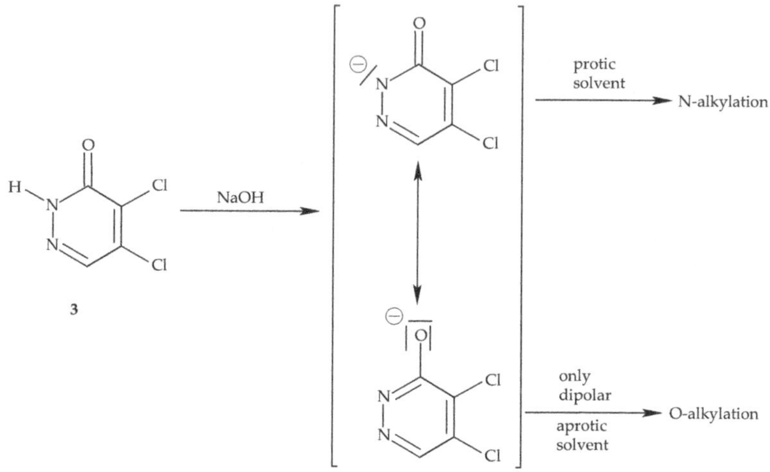

Reaction Equation 3.4 Behavior of dichloropyridazinone as an ambident nucleophile.

Thus, after the first couple successful experiments (see experiment numbers 8 and 9, in Table 3.4) we tried to minimize the amount of DMF (see experiment numbers 10–14, in Table 3.4) in order to get the most attachment with N-alkylation and the least attachment through the lactame oxygen.

We have got experimental proof that if the alkylation reaction runs in the presence of 46.2% of DMF, double alkylation happens; meanwhile the mass of the dry polymer increases to 141% of the theoretical mass of the monoalkylation. However, this extra (second) pyridazinone ring was not attached to the polymer by a stable covalent bond, since it was immediately cleaved by the attack of any nucleophiles. Consequently, we supposed that this second pyridazinone ring was attached to the polymer by O-alkylation; if the ratio of the DMF in the reaction mixture was decreased, the mass increase was much less remarkable.

Chart 3.6 shows the experimental data of mass changes of the dry polymers in the function of the ratio of DMF: we could reach 100% monoalkylation (C-alkylation) if not more than 20%–25% of DMF was added to the reaction mixture. While polymer **4** had greater reactivity, and thus only DMF of a low ratio was added, in the case of polymer **6**, we could test the whole range. The latter curve proved to be more or less a Gaussian one; pay attention only to the parts of these curves above 100% mass increase (above the horizontal violet line). It could be stated that while there was O-alkylation above 12% of DMF in the reaction mixture for polymer **4**, this limit value for polymer **6** was 25% of DMF.

The attachment through the nitrogen resulted in sufficient stability during the steps that followed. The relative excess of dichloropyridazinone **3** in the reaction mixture did not have significance, although the reactions with 3.36 times molar excess resulted in slightly better yields. We did not even try to attach the substrate through a phenol ether linker by simple boiling instead of water distillation.

4,5-Dichloro-3(*2H*)-pyridazinone **3** was a versatile substrate against nucleophiles, electrophiles, acids, and bases. However, the supported analogous substrate (either **5** or **7**) must not be attacked by electrophiles since the pyridazinone ring is highly deactivated. Thus, a S_E reaction would take place on the unfunctionalized aromatic rings of polystyrene. For this reason, the use of strong acids (like trifluoroacetic acid [TFA]) or electrophiles (like boron triiodide) was reserved for the cleavage reaction (see Reaction Eqs. 3.17–3.22).

3.3 Halogen Exchange Reactions

Conditions for these S_N reactions were set to provide excess in concentration for the reagent to the substrate, and that was ensured if sodium iodide was applied in molar excess 8–16 times more than the substrate. While both chlorine atoms of the supported substrate may react, regioselectivity toward the 5-substituted derivatives was increased by carrying out the iodination reaction under protic conditions (see Reaction Eq. 3.5). Under these conditions, it is known that the 5-iodo derivative is the main product in soluble synthesis. The 5-iodo derivatives **9** and **10** had a considerably more reactive iodine atom than the 4-chlorine atom toward S_N reactions. For preparation of polymer **10**, more vigorous conditions were needed, starting from polymer **7** (see Reaction Eq. 3.6).

The purpose of iodination reaction was dual: to increase reactivity of the leaving group in position 5 and to remove any O-alkylated substrate (if that was formed during the previous step). Therefore, the mass increase during the iodination reaction could not be used for checking the course of the iodination, since sometimes there was a bigger mass decrease due to the removal of the O-alkylated substrate than the mass increase that happened due to the building in the iodine atom into the substrate.

About the exact procedures, see Chapter 2.

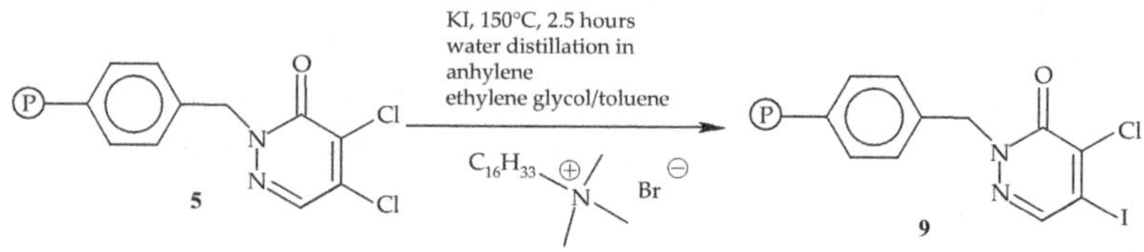

Reaction Equation 3.5 Replacement of Cl by I of polymer-supported pyridazinone **5.**

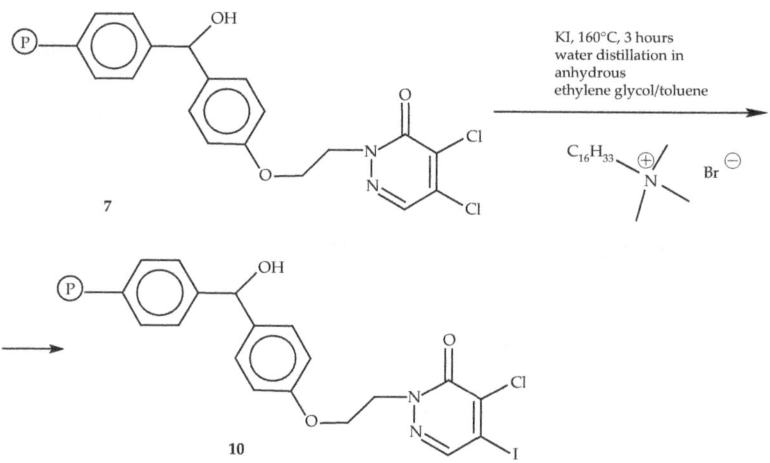

Reaction Equation 3.6 Replacement of Cl by I of polymer-supported pyridazinone **7.**

The constant parameters were as follows: potassium iodide was the reagent, HTMAB was the catalyst, the molar ratio of the catalyst to the substrate was 0.58 times, anhydrous EG was usually the solvent, and the temperature was 140°C–160°C.

Loading of the starting material was 1.41 mmol/g (polymer **5**) or 0.73 or 1.2 mmol/g (polymer **7**) (see Reaction Eqs. 3.5 and 3.6). Changing factors are summarized in Table 3.5.

Table 3.5 Iodination reactions with polymer-supported substrates

A	B	C	D	E	F	G	H
15	**5**	6.5	1.37	butanol:toluene:water	120	9	29
16	**5**	6.75	2.3	butanol:toluene:water	100	5.5	28.2
17	**5**	10.2	5.3	anh. EG	140	5	37.8
18	**5**	5.1	2.66	anh. EG	140	5	14.9
19	**5***	15.6	8.1	anh. EG	140	5	94.3
20	**5**	43.4	30.4	anh. EG	140	5	89.9
21	**5**	16.7	6.6	anh. EG	140	5	69.1
22	**7**	39.4	30.2	anh. EG	150	9	28.1
23	**7**	39.3	4.28	anh. EG:toluene	155–160	3**	62.5
24	**7**	39.4	5.45	anh. EG:toluene	150	4**	66.2

A: Experiment number
B: Starting material
C: Molar ratio of the reagent to the substrate, times
D: Relative concentration of the reagent to the substrate
E: Solvent
F: Temperature in °C
G: Reaction time in hours
H: Yield of the given step in % (based on the decrease of the carbonyl area of the IDS of the reaction, using the νC–C band of 1034 cm^{-1} as the reference band
*Starting with the polymer isolated in the previous experiment, previous row
**Boiling with water distillation

The iodination reaction was especially slow in aqueous solutions (see experiment numbers 15 and 16, in Table 3.5), so most of the reactions were run in anhydrous EG. It was important *not* to carry out iodination reaction in DMF since it would have resulted in the 4-iodinated product as the main product. Increasing the molar ratio of the reagent to the substrate was especially helpful in the iodination reactions. The molar amount of potassium iodide must be at least 10 times the molar amount of the substrate. There was strong dependence between concentration of the potassium iodide and the yield in the reaction (Chart 3.7): the reaction of polymer **5** was characterized by a saturation curve.

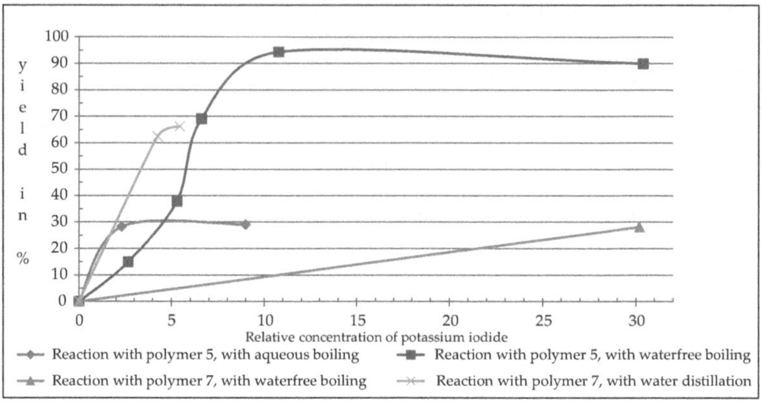

Chart 3.7 Concentration dependence of the yield in the reaction with potassium iodide.

For high yield, concentration of the reagent must be 7–10 times the concentration of the substrate. However, it was not worthwhile to increase the concentration of the reagent above 10 times the concentration of the substrate: the yield did not improve further. Probably, this was due to the high salt concentration of the reaction mixture; more potassium iodide could not be dissolved, either.

It was considerably more difficult to carry out iodination reaction with the polymer with a multidetachable linker (polymer **7**); increasing the concentration excess of potassium iodide to 10 times that of the substrate was not enough, and the reaction had to be facilitated by carrying out water distillation in the presence of anhydrous toluene:anhydrous EG mixture using a Dean–Stark trap (instead of simple boiling; see experiment number 24, in Table 3.5). The use of the Dean–Stark trap for water distillation resulted in a much greater reaction rate at a considerably lower relative reagent concentration. Thus, shortening of the reaction period became possible, from 5 hours to 2.5 hours starting from polymer **5** and from 9 hours to 3 hours starting from polymer **7**.

The temperature dependence of the iodination reaction showed 140°C–160°C as the optimum bath temperature for the iodination reaction. Great care must be taken during the iodination reaction to not omit toluene or remove toluene completely during the water distillation, since these cases would lead to the overheating of the reaction mixture, resulting in breaking of the C–O bond of the phenol ether.

3.4 Substitution by Pyrrolidine

Still the nucleophiles had to be chosen carefully to avoid powerful nucleophiles, since side reactions or cleavage did/would take place. The following reactions were run with pyrrolidine, with 2-sulfanylethylamine, and with hydrazine; and then the elimination reactions were carried out with copper(II) salts. The reaction conditions shown in Reaction Eqs. 3.7–3.9 and 3.11–3.16 were the best ones that were found; for days, none of the reactions on the inside surface of the polymer beads took place. The reaction of the polymer-supported substrate with pyrrolidine was carried out in a protic solvent mixture (Table 3.6) to ensure maximum regioselectivity toward the 5-pyrrolidino product. However, the reaction was slow at 85°C and it had to be catalyzed by HTMAB.

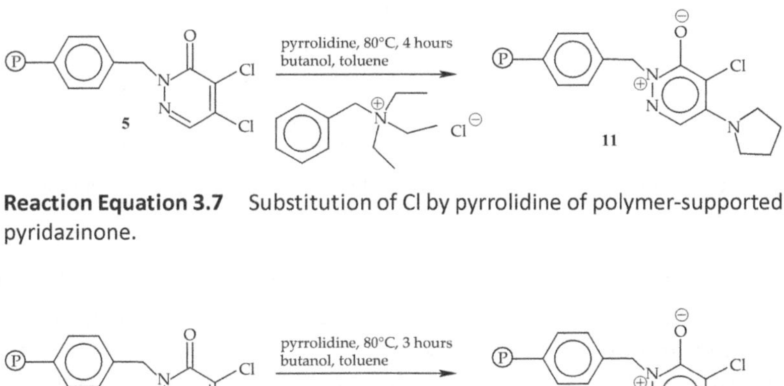

Reaction Equation 3.7 Substitution of Cl by pyrrolidine of polymer-supported pyridazinone.

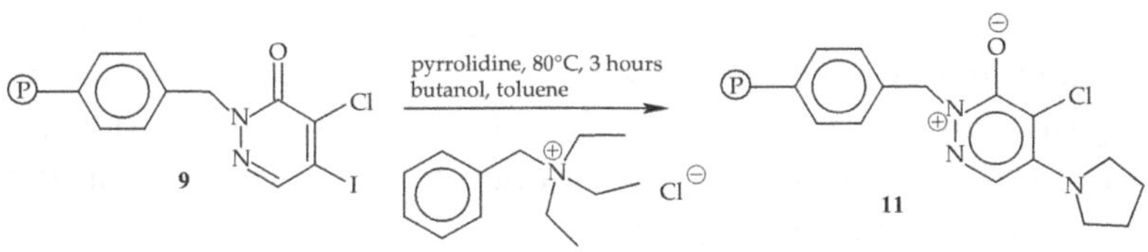

Reaction Equation 3.8 Substitution of I by pyrrolidine of polymer-supported pyridazinone.

Concentration of pyrrolidine must exceed by a considerable amount the concentration of the substrate for the reaction to complete (see Table 3.6). Chart 3.8 shows the concentration dependence of the yield in the reaction with pyrrolidine. It was clear from the weight measurements of the polymers before and after the reaction with pyrrolidine that a double reaction did not take place. There was no considerable difference between the courses of the reaction of the iodinated polymers **9** and **10** with pyrrolidine: the two concentration–yield curves were almost overlapping each other (see Chart 3.8).

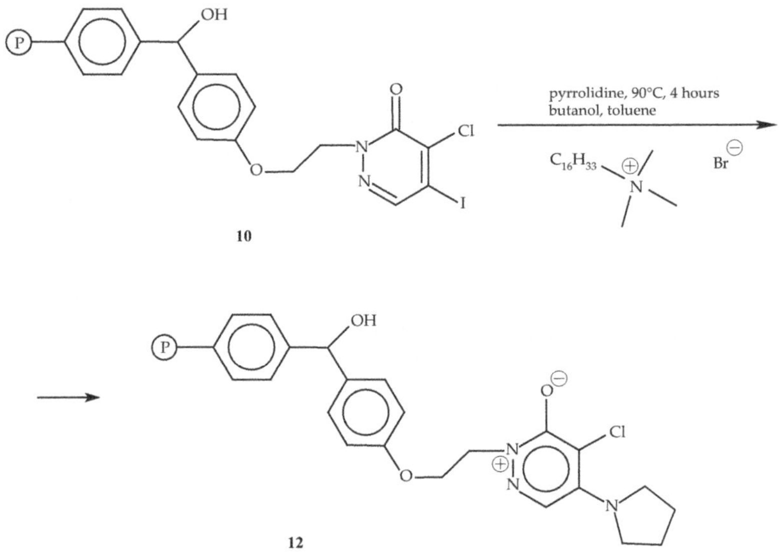

Reaction Equation 3.9 Substitution of I by pyrrolidine of polymer-supported pyridazinone.

Spectroscopic evaluation of the reactions with pyrrolidine or with other amines was not simple, since the infrared spectra of the iodides and pyrrolidino derivatives were very similar to each other (with or without standardization). For successful and reproducible electronic subtraction, the spectra had to be standardized at first (see the beginning of Chapter 2).

The constant parameters were the solvent (butanol:toluene 1:1), the catalyst (HTMAB), and the temperature of the reaction (85°C).

The ν_{arom}C=C bands, for example, the band at 1602 cm^{-1}, were usually strong, wide, and overlapping, especially in the case of the polymers with multidetachable linkers (like polymer **10**). Thus, the intensity increase at the band of 1600 cm^{-1} could not be measured or could be done so only with an increased error level. However, during the substitution reaction we observed that besides the main reaction, rearrangement of the electronic system took place: we got fully aromatic heterocycles (like polymers **11** and **12**) from structures with linear conjugation (like polymers **5**, **9**, and **10**). If aromatic heterocycles were formed, the νC=O band disappeared at 1650 cm^{-1}, and it was shifted to 1600 cm^{-1} due to the appearance of the twin-ionic aromatic structure.

Table 3.6 Optimization experiments for the reaction with pyrrolidine on polymer support

A	B	C	D	E	F	G
25	**5**	8.20x	1.84**	18.5	4	100
26	**5**	9.60x	4.65	19.4	1	52.5
27	**9**	17x	1.71	5.8	3	51.3
28	**9**	37.2x	3.42	2.6	3	84.6
29	**9**	60x	4.1	1.6	6	81.2
30	**9**	91.2x	6.82	1.1	3	96
31	**10**	5.1x	0.66	16.4	2	16.1
32	**10***	15.65x	1.87	5.4	5	+ 65.7 (totally: 81.8)
33	**10**	55.4x	5.23	1.6	5	95.3

A: Experiment number
B: Starting material
C: Molar excess of pyrrolidine to the substrate
D: Initial relative concentration of the reagent
E: Catalyst, molar % to the reagent
F: Reaction time in hours
G: Yield of substrate in the given step in %, based on the decrease of the carbonyl area of the IDS of the reaction, using the νC–C band of 1034 cm^{-1} as the reference band
*Repetition of the previous reaction with the previous end product.
**The solvent was a 1:1 mixture of 96% ethanol and toluene.

Consequently, yields were calculated from the IDS, which were created by electronic subtraction of the standardized spectrum of the iodide starting material from the standardized spectrum of the product. The νC=O areas of the IDS (with opposite signs), zeroed to the νC–C band at 1034 cm^{-1}, could be compared directly to the νC=O area of the starting material. Results were given as a percentage of the νC=O area of the starting material.

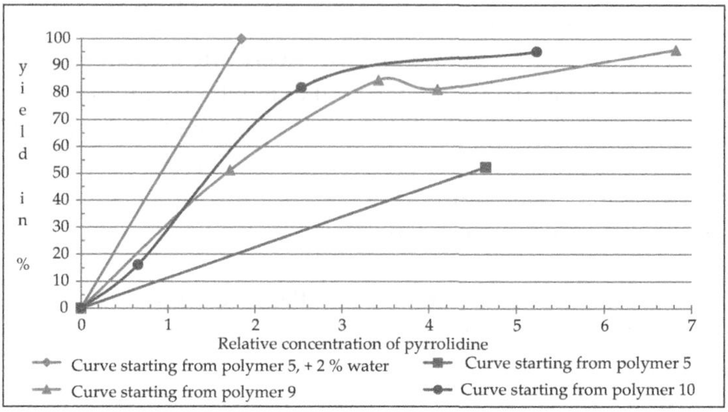

Chart 3.8 Concentration dependence of the yield in the reaction with pyrrolidine.

This kind of calculation could be used for checking yields of the reactions with pyrrolidine, hydrazine, as well as 2-sulfanylethylamine; so it looks like a general checking method for the introduction of any amine to the supported pyridazine ring: if the reaction was successful, an aromatic heterocycle was generated. But attention had to be paid *not* to add any mineral acids during the working up of the reaction mixtures or of the polymer-supported substrates. This was an important condition of the calculation mentioned in the previous paragraph; otherwise calculation would have led to false results.

While evaluating Chart 3.8, we could conclude that considerable molar excess of as well as considerable excess in concentration of pyrrolidine were needed for the complete transformation. The relative concentration of pyrrolidine was about 5 times that of the substrates with 5-iodine (polymers **9** or **10**), while about 2 times that of the simple substrate (polymer **5**) in the presence of 2% water. The presence of 2% water (see experiment number 25, in Table 3.6) was especially advantageous since the S_N1 reaction was favored by increasing the polarity/dipole moment of the solvent. Although the yields were somewhat lower under anhydrous conditions, 5-iodinated as a starting material provided greater selectivity toward the 5-pyrrolidino product than 5-chloro compound as a starting material. The catalyst amount could be decreased by about 1 order if the amount of pyrrolidine was increased.

3.5 Side Reaction with a Strong Base

With a polymer-supported substrate, only the best conditions were tried. A strong base ($NaOC_2H_5$) could not be used for deprotonation of the polymer-supported pyridazinone, since in a separate experiment, complete cleavage of the 5-monoalkylated-supported substrate took place by the formation of a thioester side product (ring cleavage of the pyridazine with fragmentation/retrocondensation took place according to Reaction Eq. 3.10).

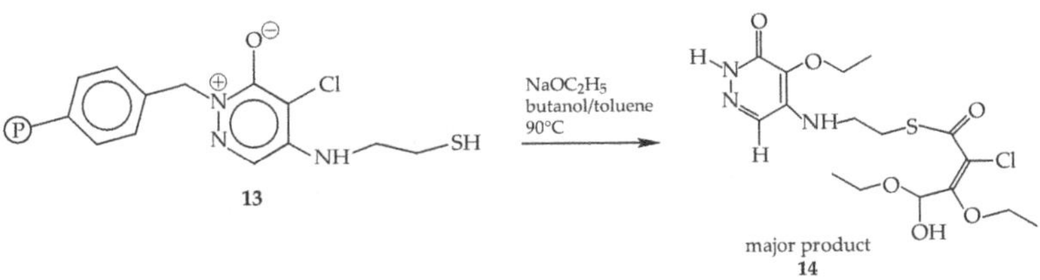

Reaction Equation 3.10 Attempted deprotonation of polymer-supported pyridazinone, resulting in cleavage.

3.6 Alkylation Experiments by Cysteamine

Thus, the following alkylation experiments were carried out with a polymer-supported substrate. The constant parameters were as follows: the reagent was 2-sulfanylethylammonium chloride, the solvent was anhydrous DMF, and the atmosphere was always argon in order to prevent any oxidative side reactions of the SH compound (the reaction was checked by weight decrease and by FTIR).

Reaction equations 3.11 and 3.12 show the changes in the reactions of the iodide:

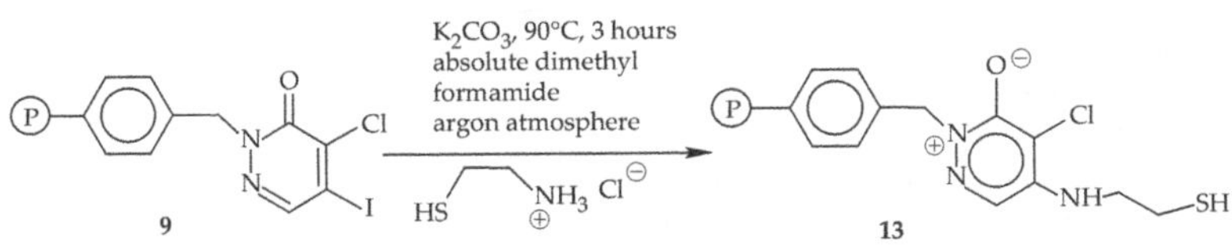

Reaction Equation 3.11 Substitution of I by 2-sulfanylethylamine of polymer-supported pyridazinone.

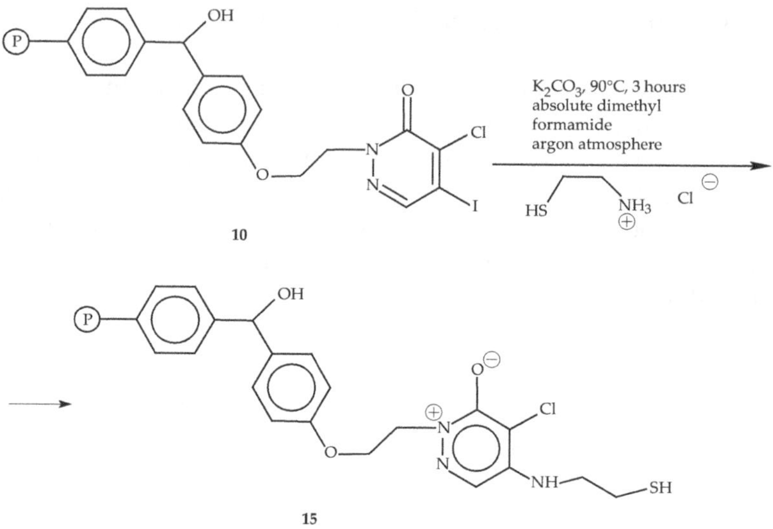

Reaction Equation 3.12 Substitution of I by 2-sulfanylethylamine of polymer-supported pyridazinone.

The changing factors are summarized in Table 3.7.

Table 3.7 Alkylation experiments with a polymer-supported substrate

A	B	C	D	E	F	G	H	I
34	**9**	$KHCO_3$	7.5x	3.7x	1.7	80	4.5	24.4
35	**9**	$NaHCO_3$	40.4x	20.2x	2.85	90	16	33.1
36	**9**	$NaHCO_3$	20.2x	10.1x	5.7	90	5	37.1
37	**9**	K_2CO_3	10.6x	5.3x	0.54	80	4	52.4
38	**9**	K_2CO_3	6x	5x	0.83	90	3	94.5
39	**10**	$NaHCO_3$	9.6x	4.8x	2.28	90	4	54.9
40	**10**	K_2CO_3	6x	5x	1.6	90	3	99.1

A: Experiment number
B: Starting material
C: Base
D: Excess of the base to the substrate
E: Excess of the reagent to the substrate
F: Initial relative concentration of the reagent
G: Temperature in °C
H: Reaction time in hours
I: Yield for products **13** or **15** in %, based on the electronic area determinations of the IDS (IR; see the text)

The alkylation reaction with sodium hydrogen carbonate took place more slowly than the similar deprotonation with potassium carbonate. Neither low solubility of the sodium hydrogen carbonate nor the high concentration of the salts caused a problem for the alkylation reaction. However, we had difficulties with this alkylation reaction since increasing molar excess of the base ($KHCO_3$ or $NaHCO_3$) as well as of the reagent strongly decreased the yield. Repetition of the reaction with more reagent and a longer reaction period did not result in complete transformation either.

Thus, it was decided not to apply a large excess of the base as well as of the reagent but to decrease the volume of the solvent in order to reach a sufficient relative concentration of the reagent to the substrate. Because of the moderate yields, we had to return to the deprotonation by K_2CO_3, and thus complete transformation was reached with both linkers (polymers **9** and **10**).

Evaluation of the reactions with 2-sulfanylethylamine was also carried out by subtraction of the νC=O areas of the standardized IDS (with opposite signs) of the starting material from that of the product, zeroing it to the IR band at 1034 cm^{-1}. Results were given as a percentage of the νC=O area of the starting material, and these are shown in column I in Table 3.7. The results are graphically depicted in Chart 3.9 to show a strong correlation among the actual kind of base used for deprotonation, the relative concentration of the 2-sulfanylethylamine, and the yield.

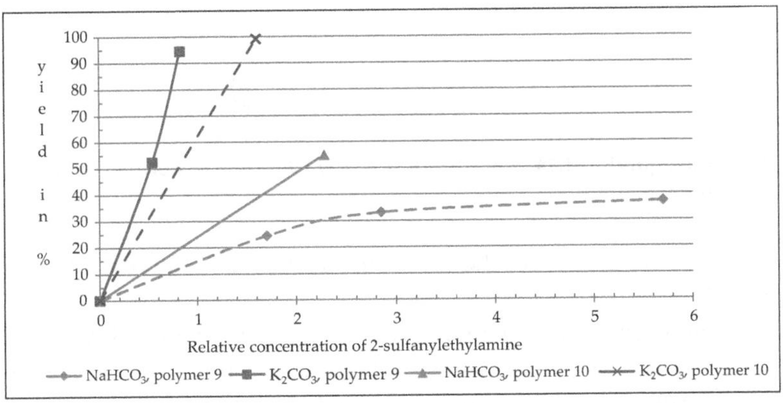

Chart 3.9 Concentration dependence of the yield in the reaction with 2-sulfanylethylammonium chloride.

3.7 Substitution Experiments by Hydrazine

Thus, the following experiments with hydrazine were carried out by a polymer-supported substrate (Reaction Eqs. 3.13 and 3.14; Table 3.8).

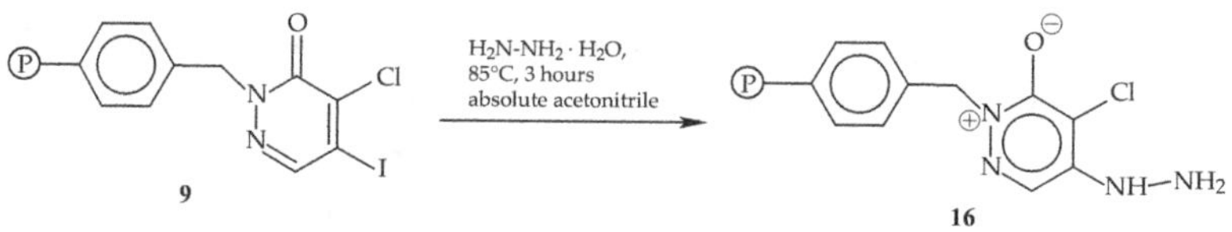

Reaction Equation 3.13 Substitution of I by hydrazine of polymer-supported pyridazinone.

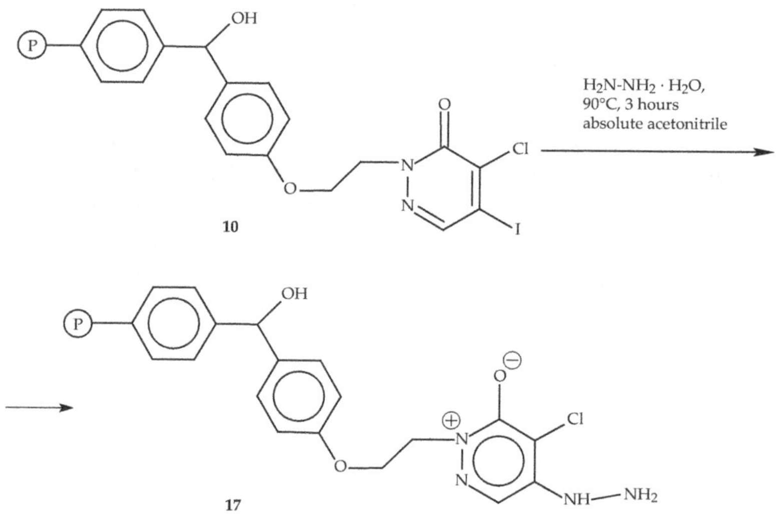

Reaction Equation 3.14 Substitution of I by hydrazine of polymer-supported pyridazinone.

The temperature of the reaction was 90°C and the reaction time was 3 hours.

The 5-iodo derivative **9** and the 5-chloro derivative **5** were of a comparable reactivity since the relative concentration–versus-yield curves (Chart 3.10) run almost parallel to each other. However, the 5-iodo derivative **10** with a multidetachable linker resulted in lower yields, especially above the relative concentration of 2, and simply increasing the relative concentration did not improve the yield due to the precipitation of the polymer. In this case, repetition of the reaction was needed (shown by the arrow in Chart 3.10). Results (yields) of Table 3.8 are graphically displayed in the function of the relative concentration of hydrazine in Chart 3.10.

Table 3.8 Optimization experiments for the reaction with hydrazine on a polymer support

A	B	C	D	E	F
41	**5**	5.35x	0.85	ACN	53.1
42	**5**	8.2x	1.27	ACN	64.3
43	**5**	11.9x	1.94	ACN	72.5
44	**5**	18.1x	2.56	ACN	73.9
45	**9**	3.6x	0.59	ACN	37.8
46	**9**	5.53x	1.06	ACN	41.7
47	**9**	7.2x	1.05	ACN	41.5
48	**9**	7.1x	1.15	ACN	61
49	**9**	14.4x	2.11	ACN	67.3
50	**9**	21.6x	2.95	ACN	71.6
51	**9**	7.2x	1.44	DMF	37.7
52	**7**	25.5x	6.36	ACN	88.2
53	**7**	30x	5.5	DMF	58.2
54	**10**	6.66x	1.1	ACN	53.9
55	**10**	13.3x	2.17	ACN	46
56	**10***	17x	4.45	ACN	73
57	**10***	27.3x	10.1	ACN	94
58	**10**	22.2x	9.6	DMF	10

A: Experiment number
B: Starting material
C: Molar excess of hydrazine to the substrate
D: Initial relative concentration of the reagent
E: Solvent
F: Yield of the substrate in the given step in %, based on the electronic area determinations of the IDS (IR; see the text)
*Repetition of the previous reaction with hydrazine, by the previous product

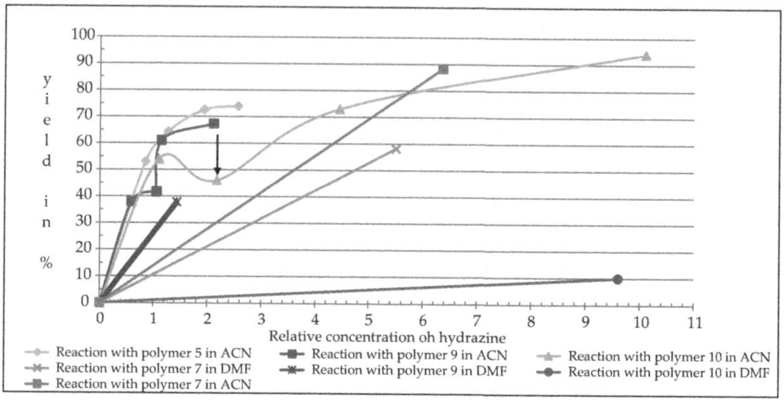

Chart 3.10 Concentration dependence of the yield in the reaction with hydrazine.

The black arrow means repetition of the reaction with the previously isolated polymer, using conditions similar to the ones before.

Shrinkage/salt formation (precipitation) happened on application of hydrazine in a high relative concentration. Another solvent (DMF) was also tried in order not to waste so much of the hydrazine. In DMF, there was no precipitation, and we expected to get complete transformation with polymer **10**. However, the measured yield was very poor (10%; see experiment number 58, in Table 3.8), with considerable darkening of the polymer, and this low yield could not be due to the precipitation of the polymer.

Then the same reaction conditions were repeated with polymer **9** as well as with polymer **7** in DMF in order to prove that this low yield was not due to experimental error. As in the case of polymer **10**, in these cases lower yield could be measured with darkening of the polymer. So, it was concluded that preparation of polymer **17** could be completed by repetition of the hydrazinolysis only, the large excess of hydrazine could not be avoided, and precipitation of the polymer did not stop the reaction.

Evaluation of the reactions with hydrazine was also carried out by subtracting the νC=O areas of the standardized IDS (with opposite signs) of the starting material from that of the product, zeroing it to the IR band at 818 cm^{-1}. Results were given as a percentage of the νC=O area of the starting material, and these are shown at Table 3.8 in the function of the concentration excess of hydrazine for each reaction. However, there was a problem with the infrared analysis of the hydrazino derivatives with multidetachable linkers, since the method applied for the quantitative determination of the yield did not work (only too low numbers could be calculated for the yield). Thus, the transformations of polymer **10** were characterized by the area ratio νN–H/νC=O of the difference spectra instead of measuring any changes of the νC=O band exclusively. The IDS were made by zeroing to the band at 1034 cm^{-1}, since it did not provide greater values than 100% (if zeroing was made to the band at 818 cm^{-1}, frequently values greater than 100% could be calculated). The intensity and area of the νN–H band were strongly influenced by the relative excess of hydrazine. The area ratio νN–H/νC=O of the IDS of polymer **10** was subtracted from the νN–H/νC=O area ratio of the difference spectrum of the reaction with hydrazine and was compared to the similar area ratio of the reactions with either polymer **5** or polymer **9**. The results, in percent, of Table 3.8 and Chart 3.10 were calculated and were depicted. However, this was quite a complicated method of analysis, and we did not want to apply this generally for the other reactions on polymer support (just for the preparation of polymer **17**), since the νN–H band was overlapping with the νO–H band (water/alcohol content of the samples highly influenced the area between 3500 and 3100 cm^{-1}; that was why spectra of dried samples were recorded only). However, the uncertainty of the calculated data was increased.

3.8 Elimination Experiments of the Hydrazino Derivatives

Cleavage reactions by nitrogen also proved that the complicated calculation method mentioned in the previous paragraph resulted in real yields and real loadings. For that purpose, at first, an elimination reaction with polymer **17** was carried out (see later) by copper(II) salts (see Reaction Eq. 3.16), then a cleavage reaction of polymer **19** was carried out with TFA, resulting in polymer **25** and the cleaved substrate 26 (see Reaction Eq. 3.20). The isolated yield of the cleavage reaction was approximately 80%, and the isolated amount of the substrate was about 3.9 mmol; consequently, the actual loading of polymer **19** must have been around 1.8 mmol/g.

Reaction Equations 3.15 and 3.16 show the transformations of the elimination reactions.

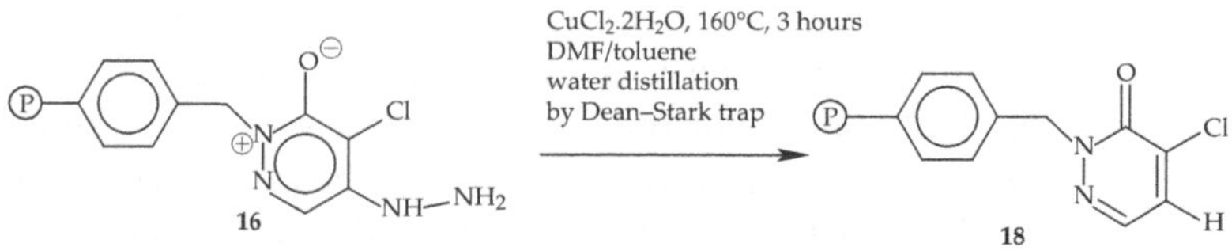

Reaction Equation 3.15 Substitution of hydrazine by H in polymer-supported pyridazinone.

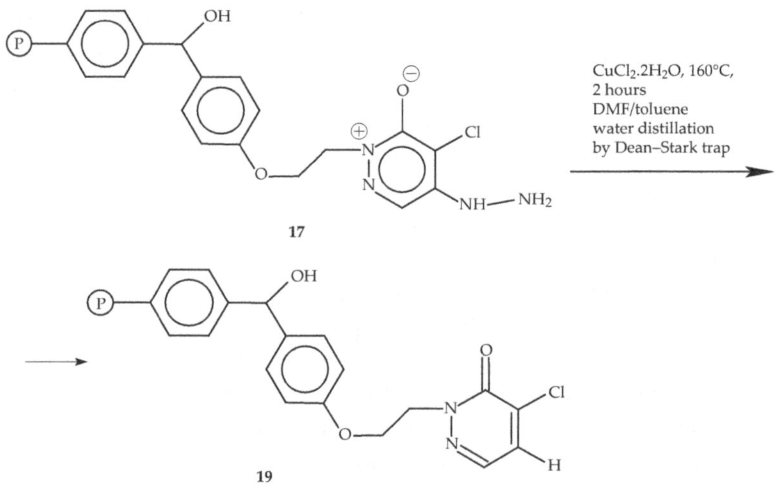

Reaction Equation 3.16 Substitution of hydrazine by H in polymer-supported pyridazinone.

The elimination reaction was preferably carried out by boiling for 1 hour and then by applying the Dean–Stark trap for water distillation. Experiments with boiling only for the elimination reaction of polymer **17** resulted in low yields. This precipitation/shrinkage was especially considerable if copper(II) sulfate was used in the elimination reaction of polymer **17**, because of the high relative concentration needed for a reasonable yield (see experiment number 65, in Table 3.9). If precipitation/shrinkage happened, then additional boiling with water or with aqueous dimethyl sulfoxide was needed after filtration, even though yields could not be calculated from the measured changes in mass of the sample, since many samples had mass above 100% instead of mass decrease, even after aqueous boiling and prolonged drying.

Table 3.9 Optimization experiments for the reaction with copper(II) salts on a polymer support

A	B	C	D	E	F	G	H
59	**16**	$CuSO_4$	28x	4.6	butyronitrile: H_2O*	3	19.8
60	**16**	$CuSO_4$	37x	2.61	DMF:H_2O	3	32.3
61	**16**	$CuSO_4$	7.5x	0.95	DMF:diglyme	2	74.5
62	**16**	$CuSO_4$	4.4x	0.41	DMF:toluene	3	81.8
63	**16**	$CuSO_4$	13.2x	0.84	DMF: toluene: acetic acid**	2	68.1
64	**16**	$CuSO_4$	13.2x	1.21	DMF:toluene	2	94.1
65	**17**	$CuSO_4$	58.8x	10.5	DMF:toluene	3	76.3
66	**17**	$Cu(OCOCH_3)_2.2H_2O$	16.5x	1.81	DMF:toluene	1.5	13.4
67	**16**	$Cu(OCOCH_3)_2.2H_2O$	10.8x	0.27	DMF:toluene	3	36.2
68	**16**	$Cu(OCOCH_3)_2.2H_2O$	21.5x	0.65	DMF:toluene	2	0 (100)***
69	**16**	$CuCl_2.2H_2O$	10.8x	0.32	DMF:toluene	1	62.5
70	**16**	$CuCl_2.2H_2O$[4*]	20x	0.76	DMF:toluene	3	did not improve
71	**17**	$CuCl_2.2H_2O$	35.1x	9.55	DMF:toluene	2	61.5 (23)[5*]
72	**17**	$CuCl_2.2H_2O$	13.5x	3.33	DMF:toluene	2	0 (70.8)[6*]

A: Experiment number
B: Starting Material
C: Type of reagent
D: Molar excess of copper(II) salt to the substrate
E: Initial relative concentration of the reagent
F: Solvent mixture
G: Reaction time in hours
H: Yield of the substrate in the given step in % (the yield in brackets is for the side product), based on the electronic area determinations of the IDS (IR; see the text)
*The polymer was precipitated.
**With 2.2 M acetic acid to the substrate.
***The desired product did not form at all; the side product was the acetate.
4*Repetition of the elimination reaction with the product isolated in the previous row.
5*The yield of the polymer-supported substrate was 61.5%, but 23% cleavage reaction took part with fragmentation.
6*The desired product did not remain attached to the polymer; complete cleavage reaction took place.

Mineral acids (HCl or H_2SO_4) could have dissolved the inorganic impurities. However, the used attachment (C–N bond) was not stable in a highly acidic medium.

Evaluation of the reactions with anhydrous copper(II) salts was also carried out by subtraction of the νC=O areas of the standardized IDS (with opposite signs) of the starting material from that of the product, zeroing it to the IR band at 818 cm^{-1}. Results were given as a percentage of the νC=O area of the starting material, and these are shown at Table 3.9 in the function of the concentration excess of copper(II) sulfate for each reaction.

During the elimination reaction, the fully aromatic heterocycles (like polymers **16** or **17**) were transformed to structures with a linear conjugation (like polymers **18** or **19**). Consequently, the νC=O band reappeared at 1650 cm^{-1}. Yields were calculated from the IDS, which were created by electronic subtraction of the spectrum of the hydrazino derivative from the spectrum of the 4-H product. For successful and reproducible electronic subtraction, the spectra had to be standardized at first (see the beginning of Chapter 2). The νC=O areas of the IDS (with opposite signs) could be compared directly to the νC=O area of the iodo derivatives. Results were given as a percentage of the νC=O area of the iodo derivatives. Results (yields) of Table 3.9 are graphically displayed in the function of the relative concentration of hydrazine in Chart 3.11 for polymer **16** and in Chart 3.12 for polymer **17**.

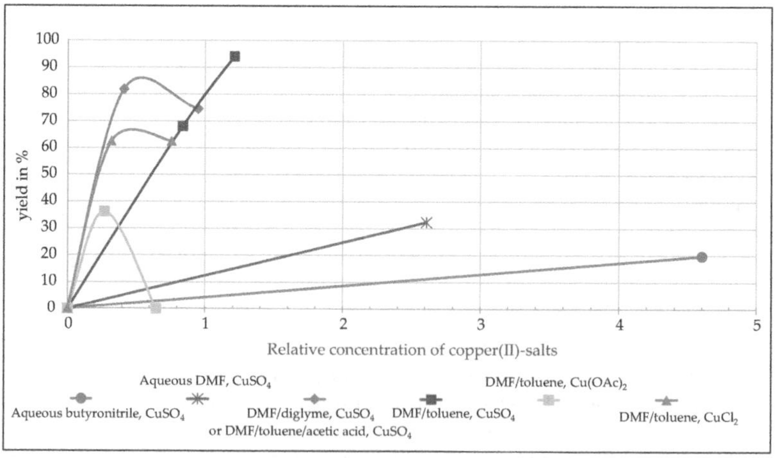

Chart 3.11 Concentration and solvent dependence of the yield in the reaction of polymer **16** with copper(II) salts.

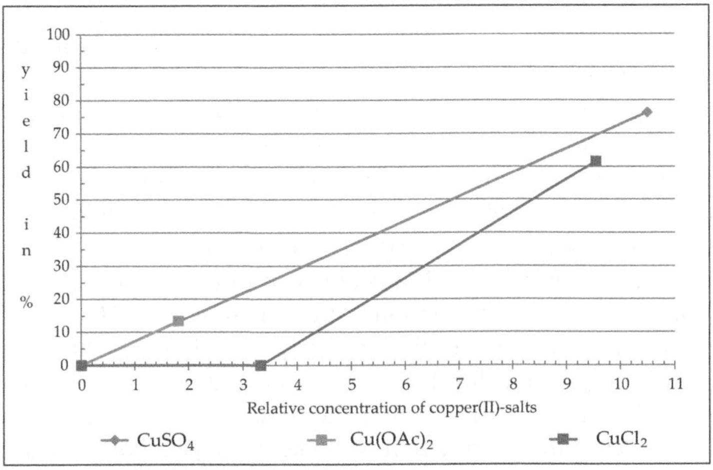

Chart 3.12 Concentration and solvent dependence of the yield in the reaction of polymer **17** with copper(II) salts.

There were experimental problems with the reaction with copper(II) sulfate. The reaction ran with considerable shrinkage of the polymer due to the precipitation of the copper(I) salts. Precipitation could not be avoided under anhydrous conditions. Many solvent mixtures were tried, but aqueous mixtures (see experiment numbers 59 and 60, in Table 3.9) resulted in extremely low yields after prolonged heating periods. Addition of a little bit more than 2 equivalents of acetic acid to the substrate decreased the yield somewhat (cf, experiment numbers 63 with 64, in Table 3.9), but isolation of this polymer was much easier, and the isolation required much less quantities of solvents.

This precipitation/shrinkage problem with copper(II) sulfate made the mass measurements useless since most of the copper(I) salts were precipitated on the large surface of the polymer beads, resulting in a large increase in mass. In this reaction, too, the pyridazinone ring was very sensitive to a nucleophilic attack. That was why no protic solvent (water, alcohol) could be present in the reaction mixture; otherwise further nucleophilic substitution reactions took place in position 4 and/or in position 5 to replace the hydrazino group. We made efforts to utilize other copper(II) salts as reagents in the elimination reaction. Going by Table 3.9, there were only two exceptions for protic solvents: acetic acid in 2.2 M to the substrate or crystalline water of the reagents ($Cu(OCOCH_3)_2.2H_2O$

or $CuCl_2.2H_2O$) could be present. The anion of the copper(II) salt had to be chosen to select as weak a nucleophile as possible. When copper(II) acetate was used as a reagent for the elimination reaction of polymer **16**; there was no precipitation, but the elimination reaction took part only partially at a low relative concentration (0.27) of the reagent. When concentration of copper(II) acetate was increased to 0.65 times, there was still no precipitation, but the polymer-supported substrate was transformed to 4-acetate completely. Thus, while 2.2 M acetic acid to the substrate was safe (did not cause acetate formation on pyridazinone), 21.5 M copper(II) acetate led to 4-acetate formation with pyridazinone exclusively.

Next, copper(II) chloride as a reagent was tried in the elimination reaction. In these experiments, there was no precipitation with polymer **16**, but the reaction stopped at a yield of 62.5%, and repetition of the elimination reaction did not help. The experiment of copper(II) chloride as a reagent in the elimination reaction with the polymer-supported substrate **17** with a multidetachable linker was very interesting. The yield was about the same as in the elimination reaction with copper(II) sulfate. However, when we tried to purify the mother liquor after filtration of the reaction mixture, we discovered that approximately 23% of the polymer-supported substrate was detached by the oxygen!

Careful analysis of the IR and ^1H nuclear magnetic resonance (NMR) spectra of the detached substrate molecule showed that fragmentation of the pyridazinone ring took place besides cleavage, according to Reaction Eq. 3.17.

It meant that copper(II) chloride behaved as a weak electrophile in this respect and partial chain cleavage happened at 160°C. On the other side, we supposed that fragmentation was caused by the vigorous conditions (excess of reagent, high temperature, and almost anhydrous solvents). Then we tried to decrease the amount of electrophile in order to get less cleavage side reaction; but applying half the amount of copper(II) chloride led to a chain cleavage in about 70% decrease of the carbonyl band, according to the analysis and area comparisons of the IR spectra of the standardized IDS, to our surprise. So it was clear that the chain cleavage did not happen by the excessive amount of copper(II) chloride, the high temperature, or the long reaction time. On the contrary, it looked like copper(II) chloride had a kind of protective role: the lower the relative concentration of copper(II) chloride, the more the cleavage reaction that took place. Decreasing the relative concentration of the copper (II) salts to one-third increased the ratio of the chain cleavage by 4.3 times. Instead of decreasing the amount (and concentration) of the electrophile, we had to try increasing the amount (and concentration) of the electrophile; but there was an upper limit value for it since more reagent could not be dissolved in that solvent mixture. Product distributions for some representative experiments belonging to Table 3.9 are depicted in Charts 3.13 and 3.14.

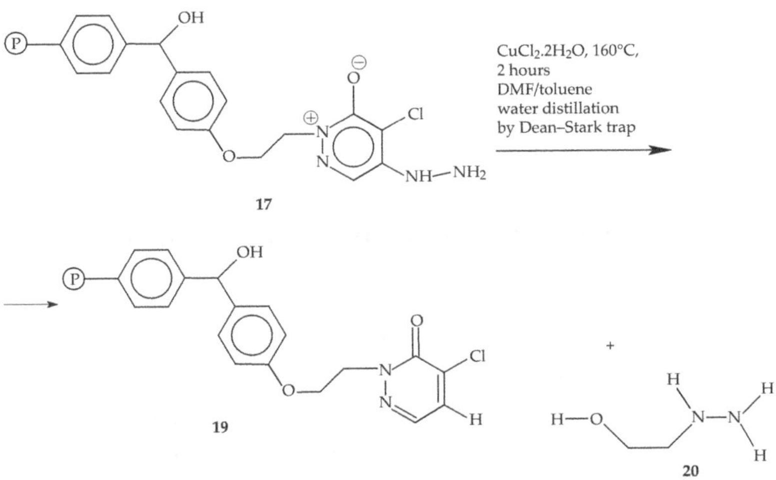

Reaction Equation 3.17 Substitution of hydrazine by H in polymer-supported pyridazinone, with a cleavage side reaction.

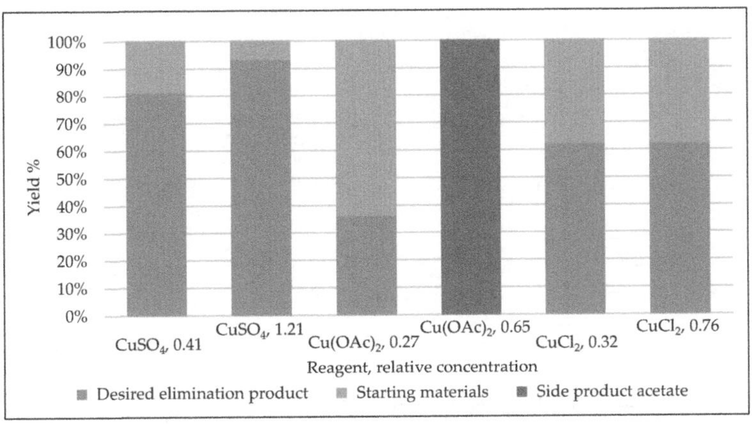

Chart 3.13 Product distribution for the elimination reaction with polymer **16**.

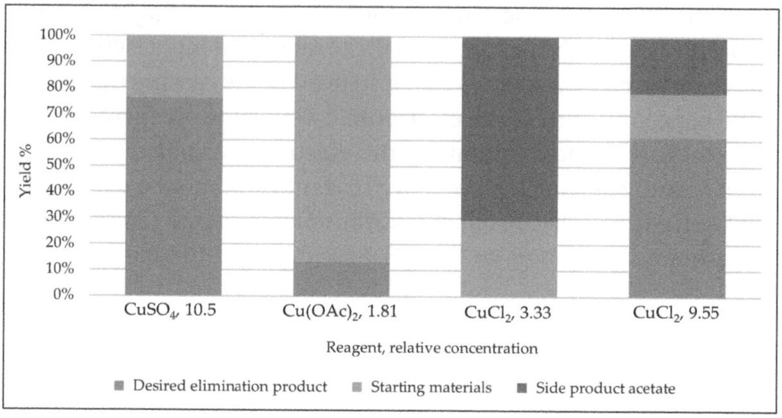

Chart 3.14 Product distribution for the elimination reaction with polymer **17.**

3.9 Cleavage of the Supported Pyridazine Derivatives

While the polymer-supported substrate **5** could be detached by an attack on the nitrogen, the similar polymer-supported substrate **7** had a multidetachable linker: cleavage could happen by attack either on the nitrogen or on the ether oxygen. The reaction equations are shown in the discussions of cleavage reactions of the specific polymer-supported substrates (see Reaction Eqs. 3.18–3.20): changing factors are summarized in Table 3.10.

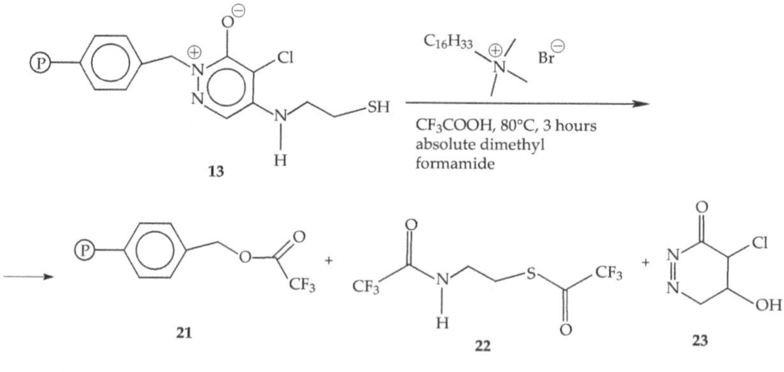

Reaction Equation 3.18 Cleavage with fragmentation.

Table 3.10 Optimization experiments for the cleavage

A	B	C	D	E	F	G	H	I	J
73	**11**	I	4.4x	1.6 (1.24)	chloroform	70	1.3	-	1.5
74	**11**	I	4.4x	2.05 (1.58)	DMF	75	2.5	-	21.4
75	**11**	I	4.4x	2.05 (1.58)	DMF	80	1.5	91	39.8
76	**11**	I	4.4x*	2.05 (1.58)	DMF	80	3.6	-	+17.6
77	**11**	I	27x	5.15 (4.67)	DMF	80	2	91	100
78	**13**	I	18.9x	3.17 (2.66)	DMF	80	3	232	58.8
79	**19**	I	4x	2.22 (1.77)	DMF	80	3	45.4	80
80	**12**	II	1.67x	0.44	DKE	100	3.5	-	48.1
81	**12**	II	5.72x	1.5	DKE	100	3.5	-	77.5
82	**10**	II	7.9x	0.57	DKE	110	2	-	32
83	**15**	II	5.72x	2.23	DKE	100	3	-	49.1
84	**15**	II	7x	1.25	DKE	95	2	-	86.7
85	**19**	II	5.8x	1.26	DKE	100	4	-	78.2
86	**19**	II	4.5x	0.81	DKE	90	3	-	48.5
87	**19**	II	8.7x	0.49	DKE	100	2	-	27.5

A: Experiment number
B: Polymer-supported substrate
C: Method I: TFA (cleavage by N); Method II: $NaBH_4/I_2$ 1:1 (cleavage by O)
D: Amount of the reagent to the substrate
E: Initial relative concentration of the reagent (real values, see the text)
F: Solvent
G: Temperature in °C
H: Reaction time in hours
I: Quaternary catalyst, molar % to the substrate
J: Yield of the cleaved substrate in the given step in %
*Starting with the polymer isolated in the previous cleavage reaction, previous row

The constant parameters for cleavage by N were cleavage by TFA and loading of the starting material, at 1.6 mmol/g (the reaction was checked by weight decrease and by FTIR, while the cleaved products had infrared, ^{1}H NMR, and ^{13}C NMR spectra after purification by column chromatography).

It was a problem of the cleavage reaction that the polymer-supported substrate usually contained a basic amino group at position 5. This secondary or tertiary amine was protonated during the cleavage; thus, the real excess of TFA was even less. That is why not only are the initial relative concentration values given in column E of Table 3.10 but also the calculated real values (in parentheses) are given, after the supposed acid-base reaction.

The substrates with 5-pyrrolidino or 5-H substituents were good intermediates for the cleavage reaction. However, the 5-hydrazino derivative blocked most of the TFA by salt formation.

Besides a similar salt formation side reaction took place with the 5-(2-sulfanylethylamino) derivative, since both N^2 and the N^5-alkyl chain had been detached, resulting in fragmentation, according to Reaction Eq. 3.18.

For this fragmentation, considerable excess of TFA had to be used, since three equivalents reacted with the substrate. Acylation on the sulfur was a side reaction, and it proved the anhydrous conditions. However, cleavage of the 5-substituent and saturation of the Δ (4-5) double bond were undesired side reactions.

Evaluation of Table 3.10 showed the importance of the relative reagent concentration to the substrate. The swelling ability of the polymers was another important factor. Chloroform proved to be a poor swelling solvent (see Table 3.10) for the polymer-supported substrate, while DMF was a moderately good (but not excellent) swelling solvent (see Table 3.10). Application of a poor solvent (chloroform) was disadvantageous because the reagent concentration was too low (especially if the salt formation with the reagent was also considered); thus, only low reaction rates were measured. When a graph was depicted from the cleavage results by the nitrogen, then two straight lines resulted (see Chart 3.15) (without a catalyst, the reaction had an especially low yield at a low (<1.5) relative concentration).

One of the curves was the result of the cleavage experiments without quaternary catalyst and the other, when quaternary catalyst was added to the reaction mixture. According to Chart 3.15, the yield was a direct function of the relative concentration of the reagent. If the latter reached the value of about 5, quantitative cleavage occurred. Quaternary salts could accelerate the cleavage reaction, and precipitation of the polymer-supported substrate did not happen.

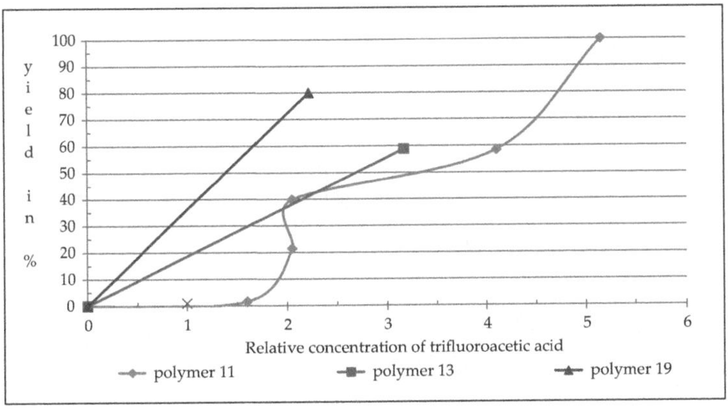

Chart 3.15 Concentration dependence of the yield of cleavage reactions.

Polymer **19** could be cleaved under milder conditions (lower relative concentration of TFA) due to the presence of the extra polar OH group as well as of the fewer NH groups to be protonated.

Reaction Equations 3.19 and 3.20 show the successful cleavage experiments by nitrogen.

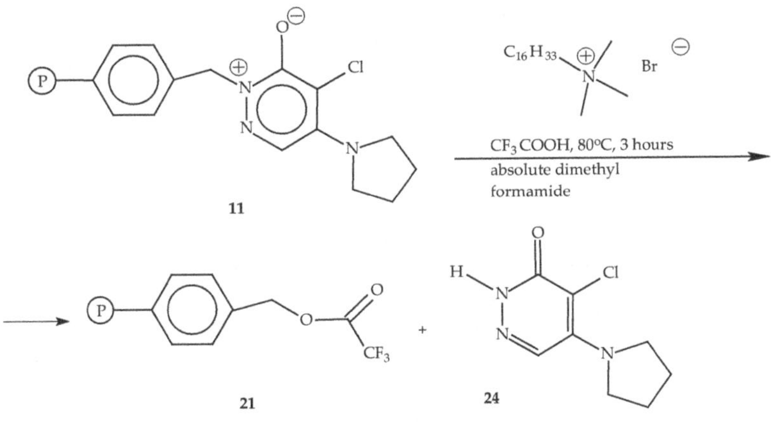

Reaction Equation 3.19 Cleavage by nitrogen.

The cleavage reaction by O was carried out by boron triiodide prepared extempore, starting from polymers **12**, **15**, or **19** and **10**. Reaction Equations 3.21–3.24 show transformations of the cleavage reactions, while the experimental data are added to Table 3.10.

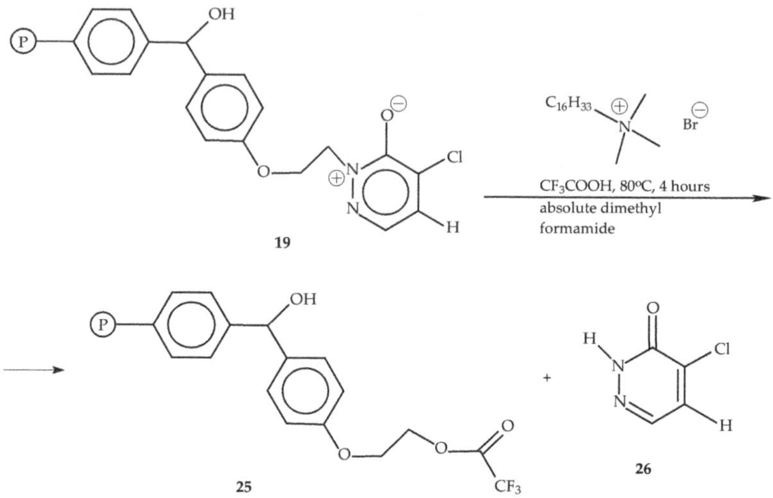

Reaction Equation 3.20 Cleavage by nitrogen.

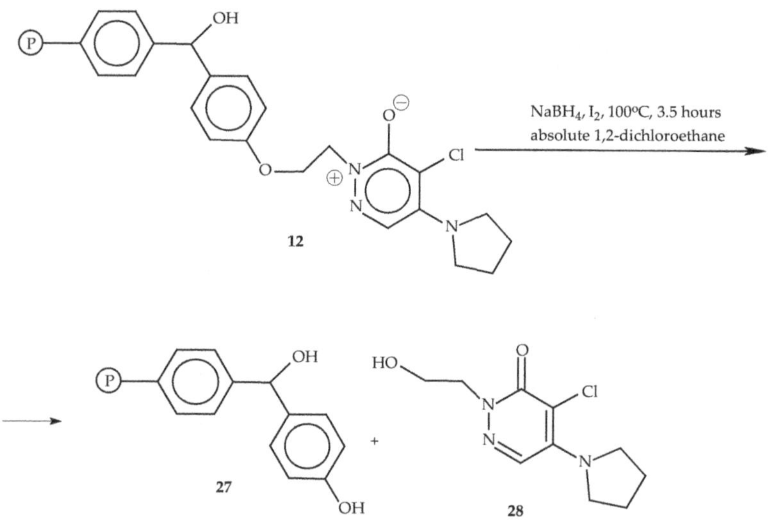

Reaction Equation 3.21 Cleavage by oxygen.

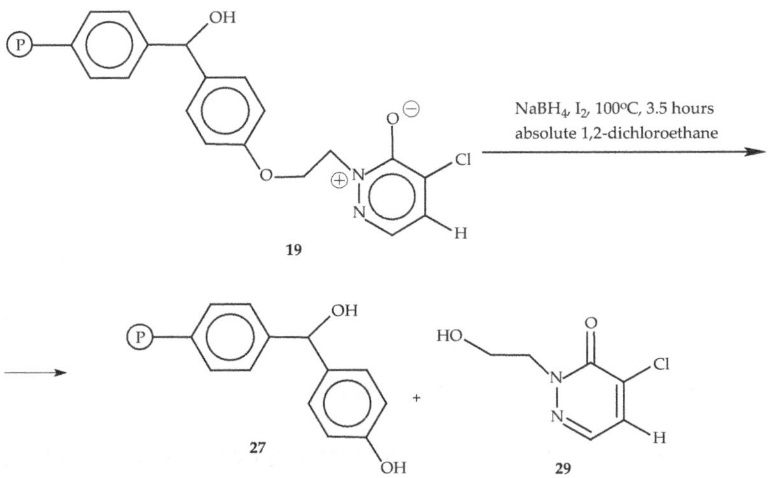

Reaction Equation 3.22 Cleavage by oxygen.

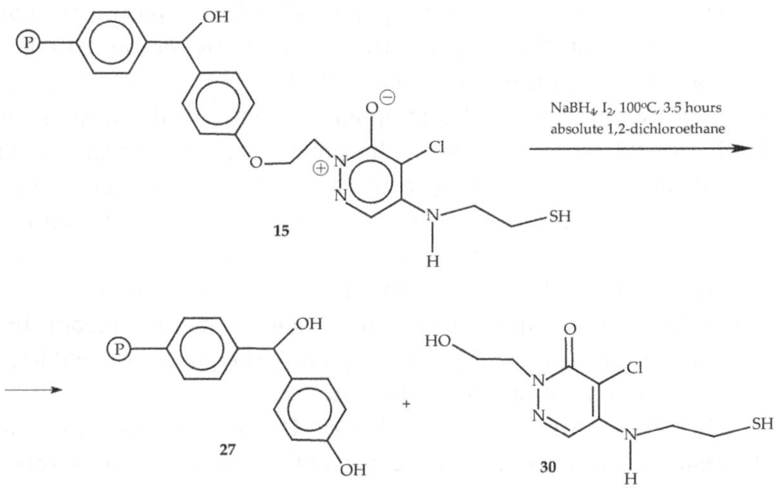

Reaction Equation 3.23 Cleavage by oxygen.

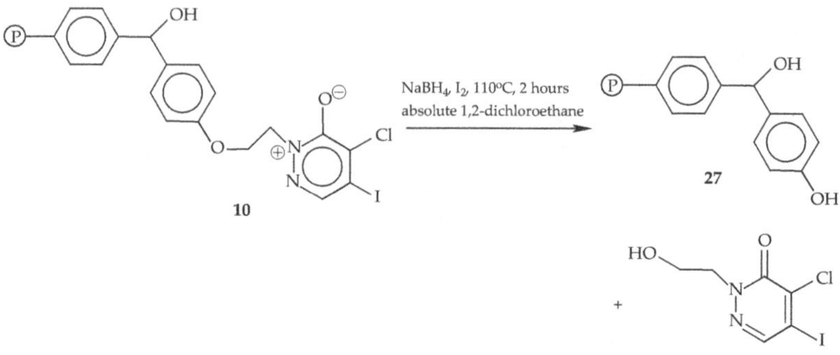

10 27 31

Reaction Equation 3.24 Cleavage by oxygen.

The reaction with boron triiodide was quicker than the cleavage reaction by TFA, and it required less excess in concentration of the reagent. Generation of boron triiodide is based on known literature data (see Ref. [13]). However, while the cleavage reaction by TFA gave yields linearly proportional with increasing relative concentration of the reagent, the cleavage by boron triiodide followed the saturation curve (see Chart 3.16). For the cleavage reaction by boron triiodide, an apolar solvent (1,2-dichloroethane [DKE]) could be used because the polar properties of the linker in polymer **12** prevented precipitation of the polymer. It was not an obvious problem at all. Polymer **12** had 6 heteroatoms (3 oxygen and 3 nitrogen), and the electrophile was coordinated to each of them. The partial bonds between the boron and the heteroatoms were quite ionic, and their formation meant the formation of new cross-links. Consequently, the polymer could have precipitated more easily, but it did not.

The cleavage reaction of polymer **19** needed less excess in the relative concentration of the reagent to the substrate to reach quantitative cleavage, since there were fewer heteroatoms to be complexed by boron triiodide.

Despite the fact that the formation of boron triiodide was incomplete and unreacted sodium tetrahydrido borate was always present, we did not observe any saturation of the pyridazinone ring: the carbonyl band remained at 1628 cm^{-1}. Further nucleophilic substitution of the terminal OH group did not happen since the electrophile was low in concentration.

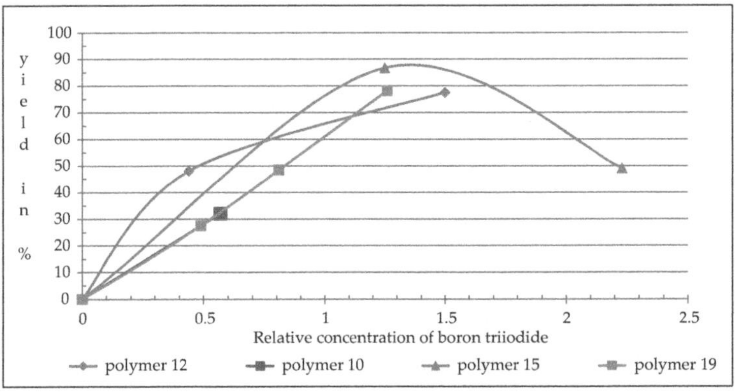

Chart 3.16 Concentration dependence of the yield of cleavage reactions.

As a general conclusion, this kind of cleavage by TFA worked well if no primary or secondary NH was present in the substrate. In other cases (different amines in the substrate), cleavage by oxygen was the only method applicable.

Chapter 4

Conclusion

The purpose of our research was to prove that the simple chemistry of pyridazinones could be built within the framework of polymer-supported synthesis, especially using our multidetachable linker. This was a new research direction, differing in many points from the similar currently popular synthesis techniques using polymer-supported reagents with substrates weakly bound to the polymer. Many new principles had to be elaborated and used for our research. These principles provide new directions for polymer-supported synthesis of nonbiopolymers. Our research could be characterized by the following points of view:

- Examples were provided for synthetic transformations of pyridazinones, using two kinds of polymer-supported linkers.
- The multidetachable linker was reusable many times; thus, it was not only accessible for anyone (easily prepared from polystyrene) but also economical.
- Ring synthesis of the pyridazinone on the polymer support was avoided since it would have mainly led to cross-linked side products .
- A stable attachment (within a pH range of 2–12 and up to temperature 170°C) was established for the N^2-alkylated polymer-supported substrate.

Polymers and Pyridazines
Péter Tétényi

ISBN 978-981-4800-47-1 (Paperback), 978-0-429-29646-8 (eBook)
www.jennystanford.com

- Further (nucleophilic) substitutions were carried out on polymer-supported substrates while the substrate still remained attached to the polymer.
- Quantitative infrared spectroscopy was used for controlling each step.
- Checking and evaluation of the courses of reactions were mainly on the basis of the area determination of the carbonyl band (ranging from 1740 to 1590 cm^{-1}) of the infrared difference spectra (IDS) of the standardized spectra; the IDS were made by adjusting zero absorbance to an appropriate reference band (e.g., to the γC–H deformation vibration at 818 cm^{-1} or to the νC–C stretching vibration at 1034 cm^{-1}), adjusting equal transmittances for the reference bands of the spectra to be compared.
- Calculations on relative concentrations of the reagents (Charts 3.4–3.16, Chapter 3) were made by using dynamic swelling values, which were obtained during separate swelling experiments (Charts 3.1–3.3, Chapter 3) in various solvents.
- Cleavage experiments were successfully carried out in the presence of acids (trifluoroacetic acid) or Lewis acids (BI_3 or $CuCl_2$), although occasional additional saturation of the pyridazinone ring took place.
- Importance of the calculations of relative concentrations of the reagents to the substrate was thoroughly proven. There was a strong correlation between the relative concentration and the yield. Meanwhile, it was also proven that the overall weight or molar excess of the reagent was not important in itself.

Our research represents a new line in the series of experiments leading toward the development of polymer supports with extreme stability as well as high versatility. We wanted to initiate a new turn in the history of this relatively fresh scientific area (there was no related publication before 1999).

Bibliography

1. Chen, Y, Lam, Y, Lee, SY. Solid phase synthesis of pyridazine derivatives using polymer-bound sodium benzenesulfinate. *Chem. Lett.* (2001) **2001,** 274–275.

2. Gouault, N, Cupif, JF, Picard, S, Lecat, A, David, M. Synthesis of #-3(*2H*)-pyridazinones on Wang resin. *J. Pharm. Pharmacol.* (2001) **53,** 981–985.

3. Parrot, I, Wermuth, CG, Hibert, M. Resin-bound thiophenols as S_NAr-labile linkers: application to the solid phase synthesis of aminopyridazines. *Tetrahedr. Lett.* (1999) **40,** 7975–7978.

4. Gouault, N, Martin-Chouly, CAE, Lugnier, C, Cupif, JF, Tonnelier, A, Feger, F, Lagente, V, David, M. Solid-phase synthesis and evaluation of libraries of substituted 4,5-dihydropyridazinones as vasodilator agents. Fr. *J. Pharm. Pharmacol.* (2004) **56**(8), 1029–1037.

5. Gouault, N, Cupif, JF, Amoros, M, David, M. Expedient method for the solid-phase synthesis of some 4-substituted-4,5-dihydropyridazin-3(*2H*)-ones. UPRES-Synthese et extraction de molecules a visee therapeutique. *Journal of the Chemical Society, Perkin Transactions 1* (2002) **2002**(20), 2234–2236.

6. Sotelo, E, Ravina, E. Pyridazines. Part 35: Traceless solid phase synthesis of 4,5- and 5,6-diaryl-3(*2H*)-pyridazinones. *Synlett* (2003) **2003**(8), 1113–1116.

7. Perez R, Beryozkina T, Zbruyev OI, Haas W, Kappe CO: Traceless solid-phase synthesis of bicyclic dihydropyrimidones using multidirectional cyclization cleavage. *J. Combinat. Chem.* (2002) **4**(5), 501–510.

8. Raghavan, S, Anuradha, K. Solid-phase synthesis of heterocycles from 1,4-diketone synthons. *Synlett.* (2003), **2003**(5), 711–713.

9. Tang, J, Huang, X. Novel and efficient solid-phase synthesis of 4,6-disubstituted 4,5-dihydro-3(*2H*)-pyridazinone. *J. Chem. Res., Synopses.* (2003) **2003**(9), 599–600.

10. Piettre, SR, Baltzer, S. A new approach to the solid-phase suzuki coupling reaction. *Tetrahedr. Lett.* (1997) **38**(7), 1197–1200.

11. Salives, R, Dupas, G, Ple, N, Queguiner, G, Turck, A, George, P, Sevrin, M, Frost, J, Almario, A, Li, A. Solid-phase syntheses of 6-arylpyridazin-3(*2H*)-ones. *J. Combinat. Chem.* (2005) **7**(3), 414–420.

12. Tétényi, P, Jr. Removal of ethoxycarbonyl groups from polymer-supported esters. *React. Funct. Polym.* (2002) **50**(2), 117–124

13. Tétényi, P, Jr. Selectivity during the cleavage of polymer supported aryl alkyl ethers. *React. Funct. Polym.* (2001) **50**(1), 85–93.

14. Szabó, LF, Tétényi, P, Jr, Pecher, J. Chemistry on polymer supports. 2. Intrapolymer interaction in alkylation of C-H acids. *React. Polym.* (1989) **11**, 47–56.

15. Kuraishi, T. 4,5-Substituted pyridazines. I. *Pharm. Bull.* (1956) **4**(6), 497–799.

16. Tétényi, P, Jr.. New general principles in polymer supported synthesis. In Bregg RK, editor, *Horizons in Polymer Research*, 2006, 1–59. New York, Nova Science.

17. Fields, GB, Otteson, KM, Fields, CG, Noble, RL. The versatility in solid phase peptide synthesis. In Epton R, editor, *Innovation and Perspectives in Solid Phase Peptide Chemistry*, 1990, 241–260. Birmingham, SPCC (UK).

18. Yamashiro, D, Blake, J, Li, CH. The use of trifluoroethanol for improved coupling in solid-phase peptide synthesis. *Tetrahedr. Lett.* (1976) **18**, 1469–1472.

19. Zdravkovski, Z, Zupan, M. The role of polymer catalyst structure on conversion of alcohols and amines to esters and amides. In Epton R, editor, *Innovation and Perspectives in Solid Phase Peptide Chemistry*, 1990, 609–615. Birmingham, SPCC (UK).

Index